Great Currents of Mathematical Thought

Great Currents of Mathematical Thought

Edited by

F. Le Lionnais

in two volumes

VOLUME

II

Mathematics in the
Arts and Sciences

Translated by Charles Pinter and
Helen Kline

Dover Publications, Inc.
NEW YORK, N.Y.

Great Currents in Mathematical Thought is a
new English translation of the 1962 enlarged edition
of *Les Grands Courants de la Pensée Mathématique*.
It is published by special arrangement with the
original publisher, Librairie Scientifique et Technique, 9 rue de Médicis, Paris VI, France.

International Standard Book Number: 0-486-62724-1
Library of Congress Catalog Card Number: 74-151133

Manufactured in the United States of America
Dover Publications, Inc.
180 Varick Street
New York, N. Y. 10014

CONTENTS

Volume II was translated by Helen Kline except for Part III, Book One, and Part III, Book Two, up through paper No. 32, which were translated by Charles Pinter.

Part III, Influences

BOOK ONE, MATHEMATICS AND THE HUMAN INTELLECT, page 3

BOOK TWO, MATHEMATICS AND PHILOSOPHY, page 23

BOOK THREE, TRUTH AND REALITY, page 69

Part IV, Appendix

Great Currents of
Mathematical
Thought

PART III

Influences

Very subtle inventions are to be found in mathematics, which are able to delight inquiring minds, to further all the arts and to reduce the labor of man.

DESCARTES

The whole of modern life is, in a sense, imbued with mathematics. The everyday acts and works of men bear its stamp; even our artistic pleasures and our moral life are under its influence.

PAUL MONTEL

All the experiences of men—individual experiences and especially collective ones—have a powerful impact on mathematics. Mathematics in turn enriches every aspect of our cultural life, though its benefits are very unequally distributed. Thus a two-way flow, consisting of influences exerted and influences received, unites mathematics with the other aspects of our civilization. This exchange back and forth runs through countless channels, some wide and direct, some narrow and sinuous, forming a network too complex to describe completely. Nonetheless, in the section which follows we attempt to give the broad outlines of this geography. We shall examine the position of mathematics in relation to the human intellect, to philosophy, to the sciences, the arts and technology, and thus complete our long search to find the role of mathematics in the development of society.

Book One

MATHEMATICS
AND THE
HUMAN INTELLECT

(Psychology and Pedagogics)

Whether mathematics is an ex nihilo *creation of our thought or whether it translates the form and structure of a material reality which exists outside of man—this is a question which every one must decide for himself. Personally we are convinced that mathematics reflects an objective reality, and we believe that most of the authors of this volume recognize or tend to confirm this view. However we have refrained from taking a stand, for we do not wish to anticipate the outcome of further debate on this subject in the pages that follow. In any case, it is undeniably true that purely mental processes play a more essential role in the creation of mathematics than in that of the other sciences. While mathematics does reflect the outside world, it also more or less distorts the picture which it presents. Thus the relationship between mathematics and the human intellect deserved to be considered as a separate topic before the other articles which follow. However, the size of this book does not allow us to devote to this matter as much space as it deserves.*[1]

---

[1] We hope to devote a second collection [never published] to a discussion of the following topics: the inner life of the mathematician; invention in mathematics; mathematics and mysticism; the appearance of mathematical ideas in children; mathematical understanding in animals; preconceived notions about mathematics; the place of mathematics in the curriculum of the future, etc.

Rarely has an opinion been so widely misinterpreted as Pascal's statement concerning what he calls the "geometric intelligence" [esprit de géométrie] and the "subtle intelligence" [esprit de finesse]. The first of these two terms is not to be taken literally,[2] but is to be understood in the more general sense of rigorous intelligence. Pascal's work in mathematics and physics is imbued as much with subtlety as with rigor; we may therefore infer that Pascal endowed mathematicians with both kinds of intelligence. An extremely flexible imagination, an awareness of the finest distinctions, a delicate subtlety of mind—are these not the very qualities which distinguish the work of many mathematicians? To be convinced that this is so, the reader need only glance through the pages of this book.

Jean Ullmo is a man with an unusually broad education. He has dealt successfully with epistemology, pure mathematics, mathematical physics and econometrics. It is thanks to this diversity, no doubt, that he has been able to reconcile geometric intelligence with subtle intelligence so fully in his mind. His article helps us to judge the thought of Pascal by placing it in its historical context, particularly vis-à-vis the pretensions of Cartesian philosophy. After this initial phase, Ullmo reviews the substance of the debate in the light of the subsequent development of science, and shows us that the two terms of Pascal's antithesis can and must, in our day, unite to form a higher synthesis.

René Dugas, one of our most eminent railway engineers, has taken a special interest in the history of mechanics. We are indebted to him for an original and insightful study on the incapacity for understanding mathematics. His article gives a fair evaluation of the role and the importance of mathematics, and stresses, in a clear and engaging manner, two of its fundamental aspects: its contribution to humanism and its usefulness as an instrument of work and research. Thus he breaks the ground for the two succeeding Books, which deal, first, with mathematics and philosophy and, next, with mathematics, science and technology.

F. LL.

---

[2] We also wish to point to a different kind of ambiguity which often leads many readers astray. Formerly the word "geometry" had a wider meaning than it has today: Besides the science of space, it encompassed in fact all of mathematics, including arithmetic, algebra and analysis. Euler, Cauchy and Hermite appear as eminent *geometers* in the academic encomiums of the 18th and 19th centuries. This broader meaning is sometimes encountered even in some present-day texts and in some articles in this volume.

THE GEOMETRIC INTELLIGENCE AND
THE SUBTLE INTELLIGENCE

by Jean Ullmo

LECTURER AT THE ECOLE POLYTECHNIQUE

THE STATUS OF THE ARGUMENT TODAY

FEW writings in our literature have been more influential intellectually than the famous passage in which Pascal distinguished between the geometric intelligence and the subtle intelligence. In generation after generation of Frenchmen, every thinking person spontaneously embraced one or the other of these two ways of thought; he chose his side and preferred either this way or that of grasping and understanding the outside world. What is more, each person more or less explicitly, more or less consciously gloried in his choice and scorned the other attitude. Thus the two parties came to look upon each other as rivals. This opposition, we may be sure, has been a setback for all intelligent thought, and has retarded the advent of a true scientific spirit and the realization of modern man.

We said a "scientific spirit"; what we mean, more exactly, is a spirit *enlightened* by science. For we will not touch upon the problems of research here, nor upon the scientific method; these have progressed together during the past few centuries, untouched by a priori distinctions or philosophical singularities. But now that the method is established and the success of science assured, we must expect and even demand that it will enrich the mind of modern man—arm him in his struggle with everyday problems both of thought and of action. The prodigious intellectual gymnastics that our species has used to transform the world in three centuries should now make an athlete of each individual in his apprehension and understanding of the world he lives in. It should, in any case, offer him the chance to be one. Has this in fact happened?

At the time that Pascal wrote, science was taking its first uncertain steps. With Galileo, with Pascal himself, the experimental method had just been born. Yet the first mathematical proofs, the first certainties conquered by pure thought, inebriated the faithful for whom mathematics and science appeared to be synonymous. And the very genius who called for man's intelligence to take possession of the universe—Descartes—believed that the battle had already been won and declared that he could explain all things by mathematically necessary deductions from self-evident initial facts. This prodigious champion of the geometric intelligence, of the "long chains of inference" which lead infallibly to the goal, believed that he already possessed the method for attaining all truths.

Pascal was more cautious, or more humble; he was also more of a physicist. He questioned the universal validity of this method drawn from mathematics, and thought it should be confined to a few branches of study. For all the rest, for the world of life, society, history, he did not trust Descartes's scheme of universal deduction. Pascal recognized that the science of his time was practically limited to mathematics, hence he could not expect it in its development to furnish him with any new means of getting a hold on the outside world. A new course had to be found, in order to understand as well as to act; therefore Pascal had recourse to practical intelligence—intelligence flexible enough to adapt to the complexity of the facts of experience. In his classic portrait of the subtle intelligence, he described its necessary resources of refinement, discrimination and critical skill.

Thus Pascal's diptych is a period piece; it reflects the state of science around 1650. It challenges the premature ambitions of a Descartes and is based upon the level of knowledge in the 17th century. But although this position was merely a provisional one, the successors of Pascal wished to maintain it. They believed that the dichotomy revealed by Pascal was in the nature of things, as if the human intellect is given once and for all, unable to improve itself by its own means, unable to perfect itself by its own successes.

The error was in fact a twofold one, corresponding to the two tablets of the diptych. When a Frenchman proclaims himself to be a Cartesian—this is our national petty sin—it is too often a caricature of true Cartesian philosophy that he presents. Descartes's most valuable, indispensable and permanent contribution is the impetus he gave to thought, the confidence in reason, the rejection of sterilizing authority. In the development of scientific method, he clarified the dual function of analysis which resolves complexity, and synthesis which restores diversity. On the other hand, Descartes's principle of admitting only self-evident knowledge, and the preeminent position which he accords to deduction—these are only historic accidents which can be accounted for by the primacy of mathematics—indeed, of Euclidean mathematics. His law of enumeration, which aspires to exhaust all of reality by thought, also antedates the exploration of the world by science.

Instead of discriminating, of holding on to what is still valid and relinquishing what is now obsolete, our supposed Cartesians imagine they are following Descartes when they maintain that everything can be proven, everything deduced—when they declare that logic is master of all. They affirm that our field of experience can be understood through mere analysis—resolution by discourse into its component parts—and then application to these parts of the implacable and unilinear mechanism of oral reasoning. They believe that they are imitating Descartes; but inasmuch as they fail to distinguish between reason, which searches, and logic, which resolves, they actually go beyond Descartes to join precisely those whom he opposed: the secular army of scholastic logicians who afflicted the human mind with sterility "by forever spinning facts into the deadly spider-webs of their syllogisms," as Lucien Febvre says in a fine recent book.[1]

On the other side of the diptych we have the supposed defenders of the subtle intelligence. Often they are simply unable to produce or to follow a simple mathematical argument, and are unconsciously seeking solace for their shortcoming. This is the ingenious breed of lawyers and shrewd essayists for whom no proposition can be proved but any defended—any thesis is worth its salt provided that it lends itself to dialectical virtuosity. According to this school there is no place for truth within the complexity of reality. For the would-be Cartesians, on the other hand, truth was the inevitable product of the machinery of logic.

However, the progress of science has taught us what truth really is, and how to attain the greatest possible degree of certainty. What conclusions can we draw from this lesson?

We remind the reader that it is not our intention to define the

---

[1] *Le Problème de l'Incroyance au XVIᵉ siecle* [The Problem of Unbelief in the 16th Century], in the collection "L'Evolution de l'Humanité."

spirit of scientific research: the skill to conceive and formulate hypotheses which anticipate experience, to discern and separate in phenomena the relevant factors; the competence to set up and to criticize experiments; the integrity to confront the initial hypotheses with the experimental results; the vision to build coherent systems out of hypotheses which have been confirmed and to infer general laws from the outcome of successful experiments. These activities bring into play the highest qualities of the intellect. They are, however, well known, and we will not dwell on them further. We will consider rather the results, the acquisitions of science, upon which our reasoning faculty can be brought to bear for the purpose of solving everyday problems of comprehension and adaptation.

Science teaches us essentially to try to *encompass reality in laws*, that is, to understand reality by means of laws, but with an understanding that is always advancing and perfecting itself in proportion as new laws are discovered and reality is correspondingly enriched. To justify this statement would be tantamount to restating all of the theory of knowledge.[2] Let us merely recall what a "scientific object" is in the light of modern science.

A "set of phenomena"—that is, a clearly delimited portion of our environment which can be followed, recognized and identified in time (such as a gas in a container or a piece of metal)—exhibits different laws as it is placed in different experimental situations. By an experimental situation we mean a combination of external factors and the technical conditions of the experiment (including the limits of precision). A law, then, is a relation in which an object interacts with a portion of the outside world in circumstances which may be repeated at will. Since it is a *relation*, the things with which the object relates are carefully defined by the conditions of the experiment.

Hence every law is conditional upon a certain mode of interaction; a set of laws so discovered "constitutes" the *scientific object* which becomes known to us in this way. We cannot know what a molecule is, for example, except by performing successful experiments which bring to light its relations with the rest of the universe. Our understanding of the molecule is modified and enriched every time we discover a new relation, a new law.

Suppose now that a set of phenomena which we have already

studied—that is, a known scientific object—finds itself in a new form of interaction with the outside world. How are we to predict and understand what will happen? We will, of course, endeavor to draw upon the knowledge that we have already acquired of the object, upon the laws and regularities which define it for us. Each of these laws represents what we may call a particular *mechanism of interaction* of the object with the outside world; this mechanism is strictly conditioned, inasmuch as it is brought into play only in certain well-defined circumstances. We must therefore assess the new experimental conditions as accurately as possible and view them side by side with conditions that have already been studied. As a result we will be in a position to evaluate the comparative likelihood that this or that known mechanism is operating in the new situation.[3]

Briefly, science teaches us to use known laws to guide our judgment in new situations, by evaluating the respective *probabilities* that each of various known mechanisms of interaction is currently in force.

An especially common case arises when we deal with a *statistical assemblage* of analogous scientific objects, where conditions may vary from object to object. Consider for example a gas at a uniform temperature in contact with a body which can act upon it; the gas molecules have different speeds (governed by Maxwell's distribution law) and only those which exceed a certain threshold velocity (they are called activated molecules) will react. Because the conditions vary from one object to the other, a number of mechanisms of interaction are simultaneously at work, affecting different objects with different respective probabilities. If we choose to regard the whole assemblage (the gas in our example) as a distinct scientific object, the laws of interaction it exhibits will be *statistical laws*. The expected behavior of the system in a new situation will then be expressed as a suitable combination of known statistical laws of interaction, to each of which we attach a probability determined by the conditions which prevail. Probability enters twice in this case, but we are doing essentially the same thing we did for the single object: we are comparing the conditions which prevail in the present situation with conditions which have already been studied. The fact of considering the

[3] There is always the possibility that a mechanism of interaction which is as yet unknown to us is operating in the new situation. This would further enrich our knowledge of the scientific object. But if the experiment has not been specifically conceived with this end in mind—and if the object is well known to us—then such an occurrence is most unlikely. Indeed, miracles are rare.

assemblage instead of all the individual objects is a matter of convenience only; it allows us to combine results pertaining to each molecule.

With these notions of statistics and probability, which are becoming more important every day in science, we touch upon what is no doubt the chief stumbling-block of contemporary man—the obstacle he must overcome before he can hope to achieve a higher average intelligence.[4] A sort of congenital incapacity to handle probabilistic reasoning is commonly observed. A long-standing logistical bias seems to render man unable to admit that an object can have two contradictory properties—for instance that the Frenchman can be both miserly and generous, simply because "the Frenchman" is a statistical aggregate in which conditions[5] differ from individual to individual. This flaw in our outlook may be due to the unconscious effect of a tradition of teaching or terminology based on scholastic realism. A most interesting study might indeed be made of the extent to which the average man's thinking has been influenced by Aristotle's theory of substantial forms (a theory whose very name is unknown to him). Generally speaking, any form of statistical reasoning is notoriously difficult for most people, as we may observe if we step for a few moments into a gambling hall.

Here at last we have found a direct application of mathematics to the purpose of improving the intellect. From our earlier account of the beginnings of science we may infer that mathematics, prior to the development of probability, played only an indirect role in the molding of the mind; it served as an instrument of work and a form of mental exercise. Mathematics is an excellent way of training the intellect owing to the faculties which it brings into play. It is an important instrument for discovering laws because of its power to draw all the consequences of arbitrarily given relations, its ability to generalize (and state as laws) the elementary relations which can be observed and repeated at will in our experiments. Mathematics is indispensable to science in its day-by-day development, but would seem less important when viewed merely as a part of the scientific heritage which guides and instructs our practical intelligence.

Descartes aspired to transpose mathematical reasoning directly into

---

[4] See, in this connection, the article "Chance and Physics" by Kahan. (Note by F. LL.)

[5] Among these conditions we must, of course, include the feelings and will of men. The concept of knowledge presented here does not, in any way, exclude free will. It only excludes pure contingency, which is the negation of all science (and all experience).

oral discourse, and he used the theory of proportion as his privileged example in this endeavor. However, the actual development of mathematical reasoning has not been able to justify his expectations, precisely because the mechanisms of deduction, consequence and generalization peculiar to mathematics have proven to be too complex and powerful to lend themselves to verbal transcription.

Integration, differentiation, elimination, transformation—none of these have counterparts in daily speech. To take only two simple examples, which are familiar because of their applications—we cannot express in everyday language the operation of seeking a maximum bound or an extremal function in the calculus of variations.

But the study of probability introduces a specific type of mathematical reasoning which can be directly transferred to the discursive intelligence and which gives it immediate instruction. The concepts of independent events and conditional probabilities; the study of correlation and regression; the development of ideas like the probability of causes, the law of large numbers and the law of small numbers, the most likely and least likely values of a random variable; normal laws, the mean deviation, the theory of fluctuations and errors —all of these notions, and many more, are truly enlightening to the intellect; they guide it and lend it assurance in its endeavors to gain mastery over reality.

We are now in a position to explain the two parts of Pascal's dichotomy in retrospect, by returning to his own text. Pascal was concerned with the degree of complexity of the objects which the intelligence investigates. The only "scientific objects" known in his time were mathematical objects, which are eminently "simple" because they are *assumed*; they are constructions of the human mind, hence transparent to it; they can be understood completely. It was therefore Pascal's contention that a method based upon the scientific reasoning of his time should apply only to this kind of objects— mathematical objects. In contrast physical objects are *given*[6] and not assumed; the physical world, in Pascal's eyes, was an unmapped region in which science had not yet learned to discern scientific objects permeable to reason. In our time such objects are increasingly capable of describing the reality they represent; while waiting for

---

[6] We might have said "proposed" rather than "given," in order to make it clear that we are not referring to the immediate apprehension of a given object— a concept so dear to the realists.

science to reach this stage Pascal could only suggest a provisional method to deal with facts that were alien to the mind. His method was indeed almost a prophecy of the scientific method.

Science today handles objects as complex as crystals and chromosomes. Whether we are trying to explain the so-called spontaneous mutations in chromosomes or the behavior of stocks and bonds following a currency devaluation, our method is basically the same. There is no essential difference in the way we approach these two problems, only a difference in the practical difficulties that we encounter. In the case of chromosomes, we must use the known properties of the scientific object which is designated by the word "chromosome." The laws of heredity reveal that it is made up of genes, which—in the light of their interactions with external agents—may be taken to be macro-molecules. Thus a few mechanisms of interaction suggest themselves immediately as possible explanations of mutations: the thermal agitation of macro-molecules, the radioactivity of the earth and the effect of cosmic radiation. It remains only to evaluate the respective probabilities that each of these mechanisms produces the required effect, in order to know if they are adequate to account for the phenomenon of mutations.

In the case of stocks and bonds, which involves the behavior of human investors, we must similarly seek out all the possible "reactions" —that is, the different mechanisms of interaction which may be operating. We can learn what these mechanisms are by studying the laws of economics and finance, by using history, which is a permanent record for the study of social phenomena, and, finally, by means of psychological reflection in this special case where we are dealing with human acts. After we have drawn up an inventory of all these possible mechanisms, the probability of each one must be estimated in the light of the historical, political, social, economic, financial and psychological conditions surrounding the devaluation. Here, as in our previous example, we make use of laws and evaluate likelihoods.

We may now conclude by dismissing the claims of both the promotors of a distorted version of the geometric intelligence, and the followers of an exaggerated notion of the subtle intelligence. By making clear the way in which enlightened minds work, we have found the synthesis for this outdated antithesis.

It is a mistake, and a caricature of Cartesian philosophy, to believe that certain truth can be attained by the rigid exercise of deduction.

To propose an unlikely mechanism of interaction is, for the true intelligence, as great an error as to propose an absurd one. More generally, to draw a conclusion and be satisfied with a single mechanism which has been inferred by deduction—to believe that in one cast of the net we have captured the whole of reality—this is the most serious mistake; it is the "sin of pride" that true science does not forgive.

It is a mistake, and a caricature of Pascal's idea of subtlety, to exhibit a fashionable scepticism toward truth and to hold that all positions are equally valid. Those who are scornful of scientific method, who believe in the esthetics of form but do not believe deeply in the power of reason over facts, learn that facts take their revenge.

Truth is not one, at the end of a rigid logical chain; nor is it indeterminate in all the possible directions of discourse. It is a *region*, mapped out by certain chosen axes. To recognize the axes and illuminate this region—this is to seek for what is possible and define what is probable; to apply to things that powerful agent, the lucidity and order of the human mind; this, truly, is to understand.

MATHEMATICS IN EDUCATION
AND AS A TOOL

by René Dugas

LECTURER AT THE ECOLE POLYTECHNIQUE

EDUCATIONAL reform is very much in vogue. The custodians of each discipline are anxious to delimit and defend their domain. In contests of this kind mathematics has not always had the upper hand. Mathematicians are generally looked upon as unusual beings—which indeed they are, just as poets, painters and musicians are unusual, but no more so. However, this observation often leads to the facile conclusion that it is unnecessary to train too many of them. And when voices were recently raised calling for "scientific equality," the idea was to reduce the importance of each science to the lowest level in order to enlarge the scope of classical studies.

Yet there is no reason to suppose that a mathematical education is inconsistent with humanism. A few centuries ago geometers wrote in Latin, which was their international language; and many mathematicians of our own day, who once had an inclination for literature or the classics, hesitated early in their career as to what line they ought to follow.

At any level of mathematical education, even the lowest, we must apparently distinguish between technique and culture. In the early school years it is indispensable to recite the tables and practice the four operations: the receptivity of a child's mind is amply sufficient for this. It is a good thing, said Poincaré, to cut up pies—mentally at least —in primary school before beginning the serious study of fractions. And before being initiated into true mathematical reasoning, it is preferable to wait even longer.

Intricate multi-level problems, scholastic exercises on mixtures, faucets and water-wheels—these are no more educational than reading a book of recipes. Once the skills of arithmetic have been mastered (and provided they are kept up), we should let the mathematical thinking of children lie dormant for a while. At an age when memory is so keen and assimilation so quick, this respite could be used to introduce children to modern and dead languages. Our grandfathers, who knew Latin—even if they knew nothing else—could decline *rosa* at a very tender age, when mathematics would have turned them away. Furthermore, long-term experience—the only experience which, in the final analysis, allows us to judge the value of a way of teaching —reveals that the only people who retain a lasting knowledge of foreign languages are those who came to them when very young.

After clearing rudimentary language study out of the way, it would be easy to find room in the older classes in secondary school—and without doing injury to classical studies—for the math programming that is now so painfully lacking.

Recent school programs have been designed with the purpose of re- viving the study of elementary mathematics (formerly sacrificed to the dogma of scientific equality by means of some specialized instruc- tion). Yet, when we glance through these programs, we are astounded by the elaborate precautions of all sorts which have been taken to eliminate, one by one, all the difficulties which reformers 40 years ago accepted as a necessary evil, or indeed as a useful drill. In the twelfth grade trihedrals have disappeared from the program along with half of trigonometry, which is an easy skill to learn and should be mastered at an early stage. The algebra of quadratic equations has been elimi- nated, with the result that most applications become impossible. In the "elementary mathematics" course of the final school year inverse functions were considered "too sophisticated" and were temporarily deleted. The directives are still critical of the details of programs, and advise not to take the freshness out of certain theories which most pupils will never have the occasion to use anyway. We almost get the impression that science is becoming progressively less important and requires fewer hours of study, or that the mathematical receptivity of young minds is in continuous decline.

Let us put these pedagogical considerations aside for a while and raise the level of our discussion by exploring the qualities which a mathematical education is likely to develop.

To begin with, such an education is a mental discipline—indeed, the most rigorous ever devised. There is no room in it for the approximate; consequently there is a much smaller risk than in other fields of entertaining illusions as to our own abilities. At every step we are given a convincing and lasting lesson in intellectual honesty. There have been, of course, and perhaps there still are those who profess facility in mathematics, like Condillac, but they have never succeeded and cannot succeed. There is no mathematics without effort, any more than there is "Latin without toil" or "Greek without tears."

Mathematical tasks, even very simple ones, engage the attention to such a point that the mind cannot wander, as it so often does with adolescents. Mathematics is so absorbing that budding mathematicians are not apt to daydream while they work.

Mathematics also teaches us to write better—if we are willing to concede that conciseness, lucidity and precision are qualities of style. From this formal point of view mathematics is no less valuable than classical studies. It may be objected that most great mathematicians have had the benefit of a classical education in their early years. But even when they are not deliberately striving to achieve formal elegance, their style is as firm and compact as any we can find. Everyone recognizes that Pascal is one of the foremost French writers; but it can be contended that his mathematical activity was only a stage in his creative life. On the other hand, Fermat was never considered to be a classical writer of our language, yet his correspondence is a model of its kind.

Instead of style, we might have spoken of language. The rigor and precision of mathematical language, and especially its great variety, which makes it possible to formulate the same intuitive notion in many different ways, compel us to develop a high degree of mental agility. The man of one book—in this case, the man of one symbolism—cannot be a mathematician.

Mathematics also develops the imagination, while at the same time forbidding it to wander beyond the limits of rigorous logic. Indeed, without using intuition it is impossible to solve the simplest geometric problem which is not of the type "prove so-and-so" with the result given in advance.

Finally, the esthetic value of a mathematical education is undeniable. Nobody can remain insensitive to the harmony of geometry. The word "elegance," which mathematicians use so often, reveals above all a concern for the esthetic which they place above the mere logical validation of a result. When confronted with a proof which seems barbarous to their esthetic sense, they are not at ease until they have managed to reduce it to a proof which is simpler, more direct, more suggestive—in a word, more elegant.

One final consideration, which is almost self-evident, is that mathematics trains the reasoning faculty. Not that mathematical reasoning is essentially more perfect or more complex than all other forms of reasoning. Classical analysis, at any rate, is governed by a logic which is basically Aristotelian and no different from the logic of all the other branches of science. What is important is that any reasoned argument—provided that the problem has been *correctly formulated*—is doomed to fail in mathematics if we neglect the minutest part of our premises as we move from hypothesis to conclusion.

With respect to these attributes, which are generally not contested, a mathematical education has imperfections, involves risks and harbors difficulties.

In actual practice, easier subjects must be mingled with mathematics so as to avoid putting an excessive burden on the minds of students, particularly those who are not exceptionally gifted. Mathematical studies, therefore, cannot be pursued to the exclusion of all others.

Mathematics is often reproached with being too abstruse and therefore the privilege of the small elite which professes to understand it. Certainly we cannot deny that the failure to understand mathematics is a widely prevalent phenomenon. But the fact that it is easier to diagnose and explain than the failure to understand literature, art or technology does not make it any more widespread or dangerous. All things considered, the average result of literary and classical studies is just as disappointing as that of mathematical studies.

Still another objection to a mathematical education is its very rigor, which is in such marked contrast to the contingencies and risks of everyday life. It would be dangerous, say the critics, to turn out a large number of pure logicians, so convinced of their own superiority thàt they would refuse to make concessions to the practical necessities of our century. However, this alleged rigor is more a didactic aspect than an inherent characteristic of mathematics. Let us hope, then, that mathematical educators will not exaggerate the dogmatic aspect of their science; instead, they might draw examples from the history of science (there is no need to romanticize it) to show that the only objective of rigor is "to sanction and legitimate what the intuition has discovered," as Hadamard so aptly put it.

Besides, mathematics only appears to be a privilege of the few; like all human thought, and especially thought of a symbolic nature, mathematics is condemned to fluctuate between the two poles of realism and nominalism. Diderot maintained that "pure mathematics reaches our understanding through all our senses." He is not the only man to have claimed that mathematical concepts, although they appear to be the most abstract, have their origins in the most common experience (the reader will recall John Stuart Mill's pebbles). But beyond the elementary level, mathematics does not appear to be seriously menaced by the danger of naive realism; there is no need to cite learned theories for this. The other danger—that of nominalism—is much more to be feared. A plodding student can be a slave to algorithms, though he believes he is their master merely because of his technical virtuosity. He is a nominalist without knowing it, and not a sorcerer's apprentice.

Poincaré made a great effort to spread the idea that in mathematics there is *more* than just logic. Logic of itself is sterile by definition. Although it is able to separate the wheat from the chaff, it cannot create anything new unless it is "fertilized by intuition." The problem of validity in mathematics is therefore more than a simple matter of proving non-contradiction. We must begin by asking, "What is a mathematical fact?" This question falls within the scope of psychology, of which mathematics cannot believe itself to be independent. Hence we see that mathematical concepts are as contingent, as precarious and as liable to revision as those of any other human science.

Mathematics therefore finds itself divested of a usurped title to absolute rigor and exclusively logical abstraction. This circumstance is far more auspicious than we might at first tend to believe; it explains why pure mathematics is able to progress, in the true sense of the word, by forever broadening its own concepts. By way of example, we recall how Lebesgue dispelled the apparent paradox which some critics believed to be inherent in Cantor's notion of the transfinite numbers of the second class; he did so by showing that, if such a paradox existed, it would apply analogously to the denumerably infinite set of numbers.

Let us return now to the teaching of mathematics. I recall having read somewhere, from the pen of an eminent author, that foreigners envy our "Special Mathematics" pre-university course. We may infer that the study of mathematics at the college entrance level is more rigid and exacting in France than anywhere else. We know that the main purpose of this study is to give students an excellent technique, which will allow them to take up with ease and confidence the vaster, more modern disciplines of general analysis and mathematical physics. This technique is a necessary evil, and the very fact of having made it a subject of the competitive examinations guarantees that it will be studied.

However, the "Special Mathematics" course, while retaining the technical quality which is its chief merit, can and must evolve. It should be aired a little, and deliberately oriented more and more toward analysis and even mechanics, which should take the place of the excessively long exercises in algebra and analytic geometry. This should be done for two reasons: First, at the competitive examinations there are far fewer winners than candidates; we must think of the unsuccessful candidates also, who ought to graduate from the "Special Mathematics" program with a genuine kit of useful knowledge and not merely the mental strain which even great mathematicians admit to having suffered in this course. Secondly, in view of the considerable development of general knowledge, universities and higher institutions must free their programs of those topics which could easily be covered in "Special Mathematics." Otherwise their teaching, which is seriously limited in time, may never even reach the starting-point of current scientific developments—and this may happen regardless of the basic excellence of their program. The program will then no longer fulfill its creative function, which is to place the student at the very threshold of research. This is undoubtedly one of the deeper reasons why some of these institutions, although they have not relaxed any of their strict admissions standards, are no longer pre-eminent in the scientific world as they once were.

I do not intend to discuss the unity or diversity of contemporary mathematics (including theoretical physics). I have many reservations on this question of doctrine, as well as on the "totalitarian" character of modern theories—logistical and physical—which intend to rebuild analysis and mechanics entirely, starting with their basic axioms. Nonetheless, it is undeniably true that mathematics, like every branch of modern science, obliges the investigator to specialize. This specialization exacts a penalty, namely, the disappearance of the universal mathematician who is simultaneously a geometer, an analyst, and a physicist; indeed, Poincaré was one of the last representatives of this breed.

It is easy to see, under these circumstances, what a delicate task higher education is called upon to perform. This task consists of bringing together the common fund of knowledge which is indispensable to all the different specialties, and putting it in an easily digestible form. It consists, furthermore, of maintaining the liaison among the different specialties, calling attention to the advantages to be derived if they cooperate, and defining subjects of research involving such cooperation. Thus, "seminars" have an important unifying function; without them, the different branches of technology would develop independently of each other, without harmony, without an overall plan.

But mathematics must also play, in regard to mechanics and theoretical physics, the role of a "humble servant." Teamwork among mathematicians and physicists has become a necessity. Einstein was fortunate in that Riemannian geometry and the absolute differential calculus were available to him when he was developing the general theory of relativity; but Heisenberg was indebted to two mathematicians for his success in founding matrix mechanics, which evolved from his first, intuitive extension of Bohr's correspondence principle. The axiomatic foundations of Dirac's general theory of states and observables were discovered within the framework of Hilbert spaces, which had already been studied independently of this application. In this case abstract analysis played a major unifying role by revealing that the quantum and wave interpretations are but two "cloaks" of the same use of Hilbert space. Theoretical physics gives rise to topics of research in pure mathematics and adds moment to problems already under investigation; that is why the mathematician must descend from his ivory tower and come to the aid of the physicist.

The school of pure mathematics has always been outstanding in France, and is no less so today. But, following a natural bent of the French character, the teaching of mathematics tends to be very abstract, with a great emphasis on rigor and formal elegance. Thus, in France especially, there is a growing need for the kind of teamwork we have alluded to above, and for courses in analysis which are resolutely oriented toward applied research—in physics, fluid mechanics, elasticity, electrical engineering—as is the practice in so many other countries. Mathematics for its own sake holds a peculiar fascination for many of our mathematicians, but it should not make them abandon their concern for applications. This concern, in fact, is in the spirit of the great mathematical works of the past. There is a difference, however—namely that today teamwork is almost always indispensable.

In this connection, one of our institutions of higher learning has taken the timely initiative of creating a program of applied mathematics. If our engineers, in the exercise of their profession, make so little use of the knowledge acquired (often at the expense of great effort) in their youth, if they give themselves up in most cases to empirical procedures, if they neglect even to discuss the results of their tests in the light of mathematical statistics—it is often because they lack sufficient contact with professional mathematicians or even access to books especially designed to aid them in applied research. Their mathematical training serves them daily to the extent that it has taught them method and sound reasoning, but they do not take a fuller advantage of it because they are unable to bring about on their own the indispensable fusion of abstract theory with technical reality.

Let us hope, then, that mathematicians will not be content merely to train pure intellects, but will sometimes leave their professorial chairs to participate in applied research. In this way they will help technology to become more scientific and, as a result, more efficient. Thus, mathematics will come to be not only an incomparable part of education, but also a tool which is apt to be "marvelously useful."

Book Two

MATHEMATICS AND PHILOSOPHY

It has often been said that mathematics and philosophy are closely connected. The greatest philosophers were, at the same time, eminent mathematicians: Pythagoras, Descartes, Pascal, Leibniz. Others, such as Thales, Democritus, Plato, Saint Augustine, Condorcet, Kant, Auguste Comte, Karl Marx, Husserl and many more, had a considerable mathematical education which influenced their philosophical thinking. The materialist Francis Bacon and the idealist Leibniz were as one when they wrote the following statements. Bacon said, "Mathematics should be regarded as the alphabet of all philosophy"; Leibniz wrote, "Mathematicians ought to be philosophers, no less than philosophers ought to be mathematicians."

From the point of view of philosophy, the prestige of mathematics rests above all on its use of proven methods for attaining truth. Some unknown charm seems to grant it the clarity and infallibility of formal logic, while at the same time preserving for it the pragmatism, mobility and fecundity of experimental science. Mathematics succeeds, we might say, in squaring the circle of knowledge by combining the virtues of the a priori with the advantages of the a posteriori. Such surprising endowments could not help but arouse many attempts to find an explanation. Thus, we are naturally brought to compare mathematics once again with that other, equally rigorous but almost barren, branch of classical philosophy: formal logic.

The entire apparatus of traditional deductive logic, created by Aristotle and given its definitive form by the Scholastics, has been revitalized during the last century, thanks chiefly to the work of Boole and De Morgan. As a result, the old argument on the relationship between mathematics and logic has been renewed. Marcel Boll, who has written a number of excellent books on science for the layman, is currently working with Jacques Reinhart on a volume which aims to make some of the most important and most recent results of the new scientific logic intelligible to the general reading public. The next article, which brings us the basic propositions of their book, should help to spread ideas which were unknown only yesterday but can no longer be ignored.

However, to study the interplay between mathematics and philosophy we should not restrict ourselves to these borderline matters. The three articles which follow attempt to penetrate the depths of the problem.

The nature of time is the central problem of many philosophical systems; they were built around it, clashed over it and present widely differing views on it. During the past few decades the controversy has been renewed, and mathematics has come under attack. It has been called powerless to help us understand life, thought and the flow of time. Jean Ullmo shows us, on the contrary, that mathematics is capable of the fluidity and originality which are characteristic of every process of development. His article is written in a suggestive style which reminds us of the qualities of rigor and subtlety which he examined in his preceding article. His concise statement of the facts is a necessary record which future polemics will have to take into account.

Léon Brunschvicg, whose work on mathematical philosophy is well known[1] interested himself in this anthology. I am still deeply moved by the memory of my visits to his home on Rue Irma Moreau in Aix-en-Provence. Though he was clearly aware of the dangers that hung over his head, his serenity was not in the least impaired. We spoke about the outline of this volume and the substance of the article he was planning to write for it. He proposed to describe two different views of philosophers on mathematics and its growth, corresponding to the two broad attitudes of general philosophy; the first view is an intransigent dogmatism which asserts the primacy of philosophy over mathematics; the second, an empiricism willing to let philosophy benefit from the free development of mathematics without demanding to govern it.

Events shortened the days of Léon Brunschvicg by making the conditions of his life ever more difficult and at last forcing him into hiding. Thus, he was not able to develop this thesis, which he would have enriched with his wealth of knowledge and his acute critical intelligence.[2] We thereupon asked Paul Mouy and

---

[1] See his book, *Les Etapes de la Philosophie Mathématique* (Milestones of Mathematical Philosophy).

[2] The chance discovery, in January 1948, of the manuscript which Léon

Paul Labérenne to sketch the two parts of a diptych which corresponds closely to the one Léon Brunschvicg had in mind.

Paul Mouy, who, to our great sorrow, has also recently departed from us, was particularly interested in the history and philosophy of science. His article takes the idealist point of view without, however, denying the connections between mathematics and the real world.

Paul Labérenne has chosen to write, once again, on a topic with which he is very familiar. He worked on it before the war, reconsidered it at length during the years of his captivity in Germany, then perfected it in the light of recent publications. His article, which asserts that mathematics must maintain and renew its contact with reality, leads to the consideration of the natural sciences, which form the topic of Part III, Book Three (p. 69), and the social sciences, which will be discussed at the end of this work.

F. LL.

Brunschvicg had written for us brings a welcome denial to this statement. (See Appendix, pp. 225–234.)

32

THE LOGICAL SYNTHESIS
OF THE RESULTS OF RESEARCH

by Marcel Boll and Jacques Reinhart

EVERY science can be considered from two different angles, which
overlap to some extent and are closely connected: *established science* (or
the codification of knowledge), and *developing science* (or the extension
of knowledge).

From another point of view, we can distinguish the following two
extreme varieties of science: science as a *collection* of miscellaneous
facts, roughly classified; science as a *highly structured set* of a few fundamental facts from which others can be derived by reasoning.

To the first type belong the encyclopedic compilations of the
Middle Ages (such as the bestiaries and lapidaries); to the second,
physics.

We refer to the latter variety as a "hypothetical-demonstrative"
science. Ever since Euclid's *Elements*—that ancient first draft of physics
—it has been taken as the ideal toward which all true knowledge must
strive. Whether this is the model best adapted to every branch of
knowledge is open to question. Nonetheless, it has been willingly
accepted to this day every time a discipline has reached a sufficiently
advanced stage of development.

After Euclid's remarkable achievement, which was an outstanding
manifestation of the "Greek miracle," the hypothetical-demonstrative
ideal remained for many centuries nothing more than a program.
With the exception of a few pioneers such as Archimedes, the
movement revived only in the Renaissance with the application of
mathematics to the explanation of natural phenomena. Even now,
mathematics—defined as the science of numbers—seems to many
people the only instrument capable of bringing order to the events

of the "real world." It is commonly believed that, in any field of learning, the use of numbers is the indispensable *guarantee* of scientific authenticity. This opinion may be valid to a large extent, but with many conditions and reservations.

Aside from the advantages of using numbers for the precise description of many phenomena, mathematics (as understood above) owes its success to another of its attributes: namely, that it offers schemes of reasoning that are already highly developed.

Contrary to most dictionary definitions, mathematics deals with many things besides numbers. For example, the theory of groups and topology have been developing since the 19th century without reference to the idea of magnitude. At the same time mathematicians turned their attention to demonstrative logic, and began to draw it out of the rudimentary state in which it had remained for so long. They were motivated by the fact that the backwardness of logic was an obstacle to their own research. But Boole and De Morgan could hardly have foreseen that logic, once restored to vitality, was to become an absolutely general science of which mathematics is but an extension (though a broad and important one). Thus, a new experimental science has already taken the place of mathematics in its role of universal science; logic is to mathematics precisely as mathematics is to physics.

We may now assert that established science (results) corresponds very closely to the most advanced branch (demonstrative logic) of this new discipline, inasmuch as the propositions of any hypothetical-demonstrative science are presented as *logical* consequences of initial *hypotheses*. These hypotheses do not necessarily involve magnitudes; for example, Curie's principle of symmetry can be used only by applying logic directly. Suddenly we gain a clearer perception of the scientific organization of human events, which is a vast domain where the penetration of mathematics has been timid and sporadic.

The whole edifice of logic rests upon a very simple, very concrete experimental notion. This notion is also the basis of the two subjects which logic has annexed from dismembered mathematics, namely arithmetic and probability. It is something far more fundamental than the concept of number: it is the idea of a set or *collection*.[1] The idea of a collection is one which can be immediately apprehended, and is

[1] We will use the word "collection" here so as to avoid confusion; the word "set" is currently used in mathematics in a somewhat restricted sense.

evoked by the everyday words heap, pile, flock, crowd, assembly. We may borrow from a pertinent observation by Gonseth and define logic as the physics of collections (of representatives).

As rudimentary as this physics may appear to be, it is sufficient to distinguish important concepts, which are always being used implicitly in the most familiar arguments. We will now present a few of them.

Every topic of discussion needs to be carefully delimited; this requires, first of all, that we agree to restrict our attention to a given collection called the *universe of discourse*. For example, the Euclidian coordinate plane is the universe of discourse for plane geometry, a system of atomic nuclei is the universe of discourse of microphysics, and the domain of historical facts is the universe of discourse for sociology. Many pseudo-problems—metaphysical and otherwise— have come into being merely because their universe of discourse was never clearly stated.[2]

To take an example, let us consider a very simple universe of discourse: that of the ten digits 0 to 9. In this system, the collection A of odd numbers has five members, and the collection B of prime numbers also has five members.

1. The *complement* (with respect to the universe of discourse) of the collection A of odd numbers consists of the five even numbers (we call this the collection "non-A.")

2. The *union* of the collections A and B consists of those numbers which are either odd or prime. This collection (called "A or B") has six members. Union is a more general kind of "composition" than addition in arithmetic; indeed, the latter is a special case of union, namely the union of collections which have no common member.

3. The *intersection* of collections A and B consists of those numbers which are simultaneously odd and prime. This collection (called "A and B") has four members.

These notions are sometimes used in elementary arithmetic, although they are never explicitly mentioned. Thus, suppose we consider the collections of the prime factors of the two integers 24 and 42:

$$24 = \quad 3 \times 2 \times 2 \times 2,$$
$$42 = 7 \times 3 \times 2.$$

---

[2] We call attention to a few of the paradoxes of logic itself. Russell's theory of types was designed to avoid them, but we do not even need this theory if only we observe a few elementary precautions of the kind we have mentioned above. These same precautions would also eliminate certain other paradoxes, like the well-known one of Goedel, which the theory of types is unable to resolve.

The union and intersection of these collections determine, respectively, the least common multiple (168) and the greatest common divisor (6) of the two numbers.

Union and intersection are only two examples of *compositions,* or operations on a pair of collections (taking the complement is an operation on a single collection). For every pair of collections, there are in fact 16 possible compositions, some of which are generally neglected, a mistake by reason of their connections with other important notions (subunion, with inclusion and implication; exteriorunion with externality and contrariness; extension with distinction and contradiction; bisection with identity and homology) as long as the union is connected with another kind of inclusion and with disjunction, and intersection with overlapping and with conjunction. In general, for n collections, the number of different compositions is 2 raised to the power 2^n.

When a composition operates on several collections it gives a new collection. This is analogous to the simple situation in arithmetic where, in the expression $2 + 5$, a new number is determined by two initial numbers.

However, a logical composition does not express a fact, any more than an arithmetic operation does. But when we write $2 < 5$, we are no longer writing a number, but an arithmetic fact. The following are also *facts*:

The collection C of numbers determined by the expression

$$1 + (1 \times 2 \times 3 \times 5 \times 7 \times \cdots \times p)$$

is included in the collection E of prime numbers.

The collection D of numbers determined by the expression

$$1 + (1 \times 2 \times 3 \times 4 \times 5 \times \cdots \times n)$$

has a non-empty intersection with the set E of prime numbers (that is, D is not included in the collection of non-prime numbers).

We notice, at the same time, that in order to express extremely simple mathematical facts we must use *prior* notions, which belong to logic. These notions are also experimental; a logical formula is the expression of a natural phenomenon, no less than the laws of a falling body or of the growth of trees (Pelseneer, 1935). In fact, we have already seen that collections may be used to express facts, thanks to a logical operation (called a predicate) which has nothing in common with the preceding compositions. Inclusion and identity (the mutual

inclusion of two collections) are examples of predicates; so is the statement that two sets have a non-empty intersection. Another example is
the presence of a collection in a universe of discourse, or its absence
from it (which is tantamount to affirming, respectively, the "existence"
or "non-existence" of the collection). By the way, presence simply
means that the collection has members in common with the universe
of discourse, whereas absence means that the collection is contained
in the complement of the universe of discourse.

There are two kinds of collections, closely related to each other,
which have some rather remarkable characteristics:

1. There are collections which, if they have a member in common
with another collection, are necessarily included in it.

2. Next there are collections which satisfy not only the above
property, but its converse as well.

We can easily satisfy ourselves that the first condition defines collections which contain (in the intuitive sense) at most ONE member; we
call these *objects*. Adding the second condition has the effect of cancel-
ing the words "at most"; such a collection is called a *constituent*. Until
now we have refrained from using the two words which appear in
italics in order to avoid an apparent vicious circle. The definitions of
these words, incidentally, are quite close to their meaning in everyday
speech.[3]

When an object or constituent is included in a collection, we say
that it *fits* in the collection; when it has a member in common with a
collection we say that it *belongs* to the collection. One can easily see
that these two words are synonymous for constituents, but not for
objects. Furthermore, to deny that an object belongs to a collection is
to affirm that it merely fits in the complement of the collection (that
is, either it belongs to the complement or it does not exist). Thus we

[3] Contrary to the accepted conventions for dealing with "sets" (which
Poincaré preferred to call *Mengen* because he felt that the recognized translation
[*ensembles* in French] was inadequate), a collection with a single constituent is
identical with this constituent. This usage does not introduce any new difficulties,
and is in perfect agreement with common intuition. We are also led to associate
with collections a subsidiary notion, called the "logical symbol" of the collections
which have the characteristics of constituents. This notion is confused with the
Menge of set theory, and—a little like "the word dog does not bite"—the logical
symbol of a collection does not coincide, in general, with the collection (even for
collections of only one constituent). This distinction suffices to eliminate many
pseudo-problems.

are not dealing with a simple alternative but a double alternative (as Aristotle already knew). The neglect of what we have just said is the source of a multitude of pseudo-paradoxes, many of which (strange as this may seem) are the creation of contemporary research workers.[4]

Let us picture a needle coinciding with a graduation on the scale of a measuring instrument; this is the crucial event in almost all experiments. Yet it is only a special case of an object belonging to a collection. The concept of an object belonging to a collection is a more primitive notion which will have direct applications that are still unsuspected today.

One of the fundamental viewpoints of modern logic consists in admitting that anything may be considered to be a collection; we merely need to specify its nature, that is, the universe of discourse which contains it. In particular, *facts* are collections whose universe of discourse is "that which actually happens" or, briefly, "reality." A fact is said to be "real," or rather *true*, if it is present in the universe of discourse—that is, if it has an element in common with the latter. Essentially this is nothing more than a codification of our natural habits of thought; for a fact, when it is true, exists and is present in reality. Now let us admit that reality is *one*—in other words, that it has the characteristics of a constituent; this is in perfect agreement with common sense. It follows from what we have said earlier that if a fact is true (that is, if it has a member in common with reality) then it actually coincides with reality. Reality has no member in common with the complement (or negation) of the fact in question, hence the complement is absent from reality, or untrue. Most of the properties of facts result from this monistic aspect of reality; let us note, among other things, that facts must be regarded as objects; for every fact is either non-existent or identical with its universe of discourse, which is monistic.

Facts have one very special characteristic. We have mentioned that when a composition operates on collections it produces a collection of the same kind, whereas a predicate operating on collections is a fact. We may therefore inquire what is the relationship between compositions (union, intersection, etc.) and predicates when both operate on facts. Compositions are facts because a composition preserves the

---

[4] Notice that this double alternative invalidates the classical principles of non-contradiction and of the excluded middle.

nature of the things upon which it operates, and predicates are facts by definition. Hence we find, rather unexpectedly, that each composition may be identified with a corresponding predicate and vice versa (actually, two predicates correspond to each composition). For example, the fact of two sets having a common member corresponds to the operation of intersection, inclusion corresponds to subunion, identity corresponds to bisection, and so forth.

Thus we have found—and rigorously justified—a tremendous simplification in the study of predicates operating on facts; indeed, we have shown that the study of these predicates may now be reduced to the far simpler study of compositions of collections.

We lack the space to enter into further details; but, as an example, let us mention the rather fertile notion of *permutable* predicates operating on facts. These are predicates whose overall meaning is not altered by interchanging the facts upon which they operate. Permutability is useful in a variety of problems, but does not generally appear in ordinary forms of speech.[5]

With the help of compositions and predicates operating on facts, we can define rigorously the syntax of such invariable words as *and, or, either . . . or, if . . . then,* together with verbs such as *implies, contradicts* and so forth. This enables us to give a precise description of the structure of a number of complex facts. However, these preliminary logical notions are insufficient for a more probing analysis of scientific facts.

In particular, logic is also led to establish the syntax of adverbs such as *always* and *sometimes,* together with the corresponding adjectives *all* and *some,* which are connected with the words *and* and *or* respectively.

---

[5] School geometry is governed by these forms of speech. Let us consider the elementary example of three straight lines X, Y and Z. We can prove the following:

(i) The truth of the statement "X is parallel to Y" implies that the statement "Y is perpendicular to Z" is true if and only if the statement "Z is perpendicular to X" is true.

(ii) The statement "X is not parallel to Y" implies that the statement "Y is perpendicular to Z" is false when the statement "Z is perpendicular to X" is true.

In each of the overall predicates (i) and (ii), we may permute at will the three facts in quotation marks.

However, this is not at all apparent, even in the fairly developed form given above. As to the classical statement (if two straight lines are parallel, etc.), no aspect of the logical structure can be distinguished. Permutable forms are important because once a theorem (such as the preceding ones) has been established, corollaries are immediately obtained merely by permuting the facts which enter into the theorem. The corollaries are equivalent to the theorem, though they may appear at first sight to be quite distinct, and by the more usual methods might require an involved independent proof.

These words occur very frequently—far more frequently, in fact, than the first scientific logicians had realized. It may be asserted that the study of compositions and predicates is merely a prelude to the logic of facts—a prelude which wrongly occupied the almost exclusive attention of early research workers.

We will not linger over unfortunate false "conversions" (such as "every beast drinks water" to "whatever drinks water is a beast"). But even if we disregard crude mistakes such as these, it is disappointing to notice how carelessly the words "all" and "some" are used, even by very experienced intellects. How many are there among us who can immediately find the fallacy in the scholastic anecdote about Epimenides, which is passed from generation to generation of students without adequate commentary?[6]

These considerations are at the periphery of classical logic; however, scientific logic naturally encompasses problems which are far more complicated and which cannot be unraveled without the use of a formalism whose minutest details are rigorously fixed. A simple application follows. It is the notion of what are called *modes*, which enable us to consider facts not merely from the point of view of their truth or falsity—to which we have hitherto restricted ourselves—but from other points of view as well.

Many attempts have been made, in the course of the past few decades, to abolish the traditional restriction of allowing statements to be only either true or false. Today we are able to view these early endeavors with detachment and to judge them with the aid of more satisfactory guiding principles; in retrospect, it is clear to us that they were not successful. The intuitionists Jan Lukasiewicz, Paulette Février and others proposed games of symbols with fixed rules; these games may be fairly interesting as such, but they have no correspondence with the world of experience. Thus, for example, although the "properties" of the nth negation of a fact were well known, no information was given as to the means of effectively recognizing whether

---

[6] Correct versions of this anecdote can be reduced to the following form: Epimenides, a Cretan poet, says that Cretans *always* lie. It follows that he is himself a liar; but, in that case, he was lying when he claimed that Cretans always lie—*hence he is telling the truth.*

It is not difficult to find the fallacy in this argument. It resides in the fact that the universal *contradictory* of "(all) Cretans always lie" is not "(all) Cretans always tell the truth" (which is merely its universal *contrary*), but "(some) Cretans sometimes tell the truth." This destroys the argument of the preceding paragraph.

or not a given fact is this *n*th negation. This objection may be regarded by some as insignificant, in view of a certain conception of mathematics.[7] However that may be, it is interesting to note that the questions which motivated the work we are describing can be rigorously formulated in terms of the notions introduced in the previous paragraphs.

Thus, we are led to define the *certainty* (with respect to a fact R) of a fact P as the conjunction:

[R is true]   and   [R *always* implies P].

The *absurdity* (with respect to R) of P is the certainty (with respect to R) of the negation of P. The negations of these first two modes are termed, respectively, *litigiousness* and *plausibility*. Thus the plausibility (with respect to R) of P is expressed by:

[R is true]   implies   [R and P are *sometimes* simultaneously true].

These modes are, of course, quite relative,[8] yet they are essential in order to solve certain problems which had been considered already by Aristotle. The following are a few results pertaining to modes; they are, in fact, in complete agreement with common usage:

The certainty of a fact implies its truth.
The truth of a fact implies its plausibility.

The difficulties encountered in applying the simple true–false logic to many mathematical problems arise because, in these problems, it is not the truth or falsity of a fact P which matters, but its certainty or absurdity with respect to a system R of axioms. In particular, if we admit certain axioms, it is open to question whether it can be *decided* (demonstratively) whether a given fact is true or false. The definitions given above show that for a fact P, we are not faced with a choice between two alternatives—certainty and absurdity—as would be the case for truth and falsity. Instead, a fact P is either

---

[7] This conception, we believe, disregards the fact that the object of mathematics—no matter how abstract we consider it to be—is, in the final analysis, to rejoin reality. For otherwise, as Lebesgue said on another occasion, we are working "for the gods," fashioning for them "theorems which we will never be able to understand, and reasoning on objects of which we cannot conceive." This might become a rhetoric as hollow as that of the fanciful disputes which intrigued the medieval intellectuals.

[8] They correspond to the classical modes which are, respectively, "necessity," "impossibility," "contingency" and "possibility." These terms must be rejected because they are too charged with ontological associations.

(i) certain *or* absurd—that is, decidable, or else

(ii) litigious *and* plausible—that is, undecidable

with respect to a fact R; this agrees substantially with the intuition of mathematicians who have studied the problem of decidability.

The plausibility of a fact R with respect to a fact P is itself a fact, and we may examine, for instance, its certainty with respect to a third fact S. Thus we are led to consider simple (or multiple) superpositions of modes. The most interesting case arises when the facts R, S, etc., with respect to which the successive modes are referred, coincide among themselves. In this case we can show that eight modes (the first four described above and the four that follow) are sufficient to exhaust all the possibilities.[9]

These interesting ideas can be developed still further. First, however, we will show that logic is sufficient—without borrowing from any other sources—to build the foundations of arithmetic.

A distinction can be made between numbers regarded as adjectives (for example, *three* infantrymen) and numbers regarded as nouns (for example, *three* of the infantry)—though, usually, this difference is of no great importance. A similar distinction can be made in logic between what are called, respectively, power and number (cardinal).

A *power* is a particular collection (or class) consisting of those valuable aids, logical symbols of collections. Collections whose logical symbols belong to the same power verify the same logical fact, and conversely. We define a sequence of powers by letting the power *zero* be the class of the empty collection, the power *one* be the class of all constituents, and so forth.[10] If the logical symbol of a collection belongs to the power n, we say that the collection *has* power n. (Every constituent has power one; every pair has power two; every gross has

---

[9] It can be shown in particular that—unlike simple logic where the falsity of falsity is identical with truth—the absurdity of absurdity does not coincide with certainty. On the other hand, triple absurdity is identical with simple absurdity. We call attention to the following results because they are, all in all, about the only ones which Brouwer achieved on this subject:

    (i) certainty of litigiousness = absurdity of certainty;
    (ii) certainty of plausibility = absurdity of absurdity;
    (iii) plausibility of certainty = litigiousness of litigiousness;
    (iv) plausibility of absurdity = litigiousness of plausibility.

[10] The given definitions involve only logical notions which are already known, as the definition of constituents (given earlier) illustrates.

power 144.) Equality—or equipotence[11]—of two collections is simply the identity of their powers.

A *number*, on the other hand, is a collection with a well-defined power;[12] it is called *concrete* if we indicate the nature of its constituents (for example, five polyhedra), and *abstract* otherwise.

The notion of (natural) number and its first generalizations provide us with a more acute instrument for the analysis of facts than the laconic "always" and "sometimes." For instance, we can give more information by stating:

$$R \text{ implies (with frequency } f) \text{ P,} \quad (f \neq 1)$$

than by stating merely:

R and P are sometimes simultaneously true.

In particular, by defining *probability* we add a new chapter to the study of modes: P will be said to have probability f with respect to R if:

[R is true]   and   [R implies (with frequency f) P].

Probability one is the certainty of a fact, and probability zero is its absurdity.[13] Between these two extremes, we are now able to draw certain conclusions relating to a portion of the realm of the undecidable.[14]

Logical notions, which are indispensable in the foundations of probability theory, are no less essential for its correct development. Indeed, logic is beginning to be taught as part of courses in probability theory.

The aspects of deductive logic which we have just reviewed conform to the general line of Aristotelian thought—a thought inspired

[11] We can show that *equipotent* is synonymous with "between which there exists a one-to-one correspondence." (We believe that it is preferable not to start from this latter point of view, because the concept of number is more primitive than the general concept of relation.)

[12] Number and power are very often confused with one another. The *identity* of two powers corresponds to the *equality* of two numbers. The sum of two numbers is their union (we assume they are disjoint), hence the sum contains each number. The sum of two powers is a new power which is disjoint from both of the component powers.

[13] These probabilities do not correspond, respectively, to truth and falsity, as is too often held.

[14] This is the portion in which we are given that R is true, namely, the intersection of the certainty of litigiousness with the certainty of plausibility.

by the natural sciences and attempting, in imitation of them, to order all things in great inclusive classifications. However, it may be objected against this leading idea that it seems to ignore the very notion which dominates all of mathematics: the notion of *relation*. Is it not, in fact, common knowledge that the logic of Aristotle is inoperative with respect to relations? Indeed, at one time there were grounds for wondering if mathematical reasoning was not totally alien to logic— if it did not progress by its own characteristic laws.

After having been in conflict for a long time, these two points of view have at last been reconciled in modern logic; it has, in fact, been possible to extend the method of classification to the study of relations. This too has been done by introducing auxiliary collections—a procedure to which we have had recourse once before. If A is some collection, we are led to consider auxiliary collections consisting of *pairs* of elements of A in a given order.

For example, let us consider a collection of pairs of constituents where the first constituent of each pair is the father of the second constituent. When we wish to know if X is the father of Y, we merely verify whether or not the pair X, Y belongs to our collection.

Thus, the logic of relations[15] reduces to that of collections, by considering collections of a special kind to which we can apply the results already obtained for collections in general.

Logic cannot hope to do more than define basic notions; the mathematicians, who are specialists, must then undertake to search the depths of this vast domain.

All in all, there is no longer a significant gap between deductive logic and mathematics. The mutual boundaries of these sciences are merely a matter of convention, and may be adjusted as the occasion requires.

The study of modern logic is amply justified if we merely consider the importance—epistemological as well as practical—of the new concepts it has introduced. But in addition to this, we might ponder over the extraordinary power of reasoning which the new apparatus of scientific logic has achieved. It has often been held that clarity and rigor are conflicting aims. In delicate problems rigor is required at any cost; yet logic succeeds in retaining it without sacrificing clarity, thanks to the use of a formalism which is able to bring out precisely

---

[15] Relations may be identified with predicates. Compositions—of which we have seen a few examples above—are also relations.

the structure of any problem, to furnish us with unimpeachable proofs, and to do this with an apparently disconcerting rapidity.

We cite by way of a first example a demonstration (based on that of a well-known mathematician) concerning a principle of induction.[16]

Let $P(n)$ be a statement about a number n. Suppose that, whenever this statement is true for all integers k preceding n, it is necessarily true for n. Then $P(n)$ is true for all integers.

Otherwise, the set of natural numbers would include numbers n such that $P(n)$ would be false. In particular, there would be a first number s such that $P(s)$ would be false. But then, since the statement is true for all numbers preceding s, it should be true for s also, in view of our initial hypothesis. Thus we are faced with a "contradiction." This proves that $P(n)$ is true for all natural numbers.

We leave it to the reader to use the above principle in order to prove its contrapositive, namely the statement "if there is at least one number n such that $P(n)$ is false then there is a first number for which it is false." The reader will undoubtedly be surprised to learn that, if these ideas had been expressed in the formal language of logic, then the statement and its contrapositive would be as visibly equivalent as the two expressions:

$$x - y < z \quad \text{and} \quad -z < y - x.$$

To take another, more general example, let us consider the following four statements, the usual formulations of which are so open to criticism:[17]

If two reasons are identical, then each consequence of the first is identical to a corresponding consequence of the second.

If two consequences are distinct, then some reason for the first is distinct from all the reasons for the second.

If two reasons are distinct, then some consequences of the first are distinct from all the consequences of the second.

If two consequences are identical, then every reason for the first is identical to a corresponding reason for the second.

These rules, which are used as a matter of course in every form of

[16] We might call this principle "Fermat's induction principle," because he appears to be the first mathematician to have used it, in his method of "infinite descent." The other well-known principle of induction is sometimes named for Pascal.

[17] A fact R is said to be the *reason* for a fact C (and C is said to be a *consequence* of R) if R always implies C.

reasoning, are in fact all equivalent to one another—that is, they are four aspects of a single statement. They can be deduced from one another as easily in the symbolism of logic as the two inequalities, above, can be derived from one another in arithmetic. To deduce these rules from one another without the use of logical symbolism would prove to be very difficult; yet the problem we are dealing with here is of an elemental simplicity as compared to other problems which must be solved by the methods of logic. What is more, many mathematical proofs which appear to be beyond reproach do not satisfy logicians (this happens to be the case for the proof which we gave earlier of Fermat's induction principle). If the flaws in these proofs were mended—using ordinary, verbal logic—this would lead to arguments of such interminable length as to weary any mathematical reader beyond endurance.

We have given a brief survey of some aspects of formal logic, which has become—concurrently with mathematics—the specialized instrument for organizing the major sciences. Logic enables us easily and surely to derive all the *consequences* that require the *reasons* which are furnished to us by experiment, and which we agree to take as axioms, theories, principles, etc., of *established science*. It is within the context of logic, therefore, that established science is formulated and communicated.

Scientific *research* (that is, developing science) itself has two purposes:

(i) To refine concepts—that is, to give more power and precision to our mental equipment, and to exhaust all the consequences of known facets.

(ii) To establish new facts which take their place among the preceding ones, to shed new light on known facts and sometimes to restate them—that is, to add to and reorganize established knowledge.

In connection with the first of these purposes, modern formal logic is destined to play an important role; scientists, having learned the manipulative techniques of logic, are better armed. Nonetheless—despite a popular belief in the contrary—logic is unable by itself to give a direction to research, to make useful definitions or to seek out the important results. While scientific logic is incomparably more powerful than the naive logic of high school geometry, it is nonetheless a mere instrument, just as an electronic computer[18] is basically nothing more than an abacus (raised to the level of modern technology).

Already the first type of scientific research proceeds gropingly, by trial and error—the confident rigor appearing only in finished research papers. Galois had this very aspect of science in mind when he said: "Though mathematicians may try hard to deny this, the fact is that they do not [prove their results]: they combine, they compare— and when at last they have fallen upon the truth, it is the result of fumbling, failing and searching further." The creative act in science is related to the creative act in art. In addition to a certain optimal balance between the disposition to feel and the disposition to act (which we need not go into here), the quality of the intellect which it requires above all is a very fertile imagination. This is often (but not always) accompanied by penetrating and unerring judgment. Intuition is, properly speaking, the result of imagination; and it is effective to the extent that the research worker is sufficiently educated in scientific method and informed of the results obtained in his field.

In connection with the second purpose of scientific research the situation is analogous, inasmuch as there is no "idea-making machine" and no systematic method to search for new facts. In the previous paragraph we dealt with the problem of demonstrating that a fact (believed to be true) is the consequence of certain known facts; once this logical connection has been established the scientist's work is finished, except for possible improvements of an esthetic nature. At present, however, we are dealing with the reverse of this situation; the physicist, for example, has direct knowledge of facts (the raw results of his experiments) which he considers to be consequences of a given reason which he still does not know. Drawing on his intuition he guesses what this reason is; and if he is right he will prove rigorously that the initial facts are indeed the logical consequences of this reason. But his work does not end there, because a given consequence may have a multitude of different reasons. Hence he cannot rest until he has examined all possible reasons—or at least as many as suggest themselves to him. He will establish, logically, the consequences of each one and compare them with the actual results of the experiment;

---

[18] The electronic numerical integrator and computer (E.N.I.A.C.), weighing three tons, contains 16 thousand electronic tubes and costs 400,000 dollars (Moore School, University of Pennsylvania). To fix our ideas, its use makes possible calculations four hundred thousand times more rapidly than can be done by hand.

many suggested reasons will be invalidated in this way. This process of elimination continues, sometimes for generations, until in the end a single reason remains; it is not certain to be correct, but it is nonetheless a useful summary of work up to a given time. The physicist accepts it as such, and assumes it to be valid on a provisional basis, until new experimental evidence alters the situation.

The scientist is working on a *gamble*, and it is this very form of gambling which is the subject matter of *inductive logic*. At this stage, the main function of logic is to carry out the process of elimination in such a way that the reason which is finally adopted yields very nearly the desired consequences, and no others that which stimulates notably to verify the consequences of the facts in domains of very different types. If this has been done successfully, there will remain very few extraneous consequences such as always run the risk of being contradicted by later experiments. In fact this, in the major sciences, only leads most often to limit the field of consequences of facts initially posed, which facts need then be only slightly modified.

In the major sciences, a large part of the research effort is aimed at unifying individual reasons into *theories* which encompass them all. The position of privilege which logic and mathematics enjoy in relation, say, to physics—whose basic axioms are still in the process of being adjusted—is due to the fact that they constitute a surprisingly simple process, whose basic principles were discovered early (even if this is not put to use conveniently). Nonetheless, these principles remain subject to revision, like those of any other science.

In this brief study we have attempted to show that modern deductive and inductive logic are—to an increasing degree—becoming directly involved in building other sciences. Logic is, in a sense, their cement. It bears no resemblance to that "dialectic" which a sophistical metaphysician claims can embrace all of science—even though it has not been able to account for a single correct scientific argument. Modern logic is the result of *scientists'* reflecting on the activities of science; it represents only the scientific spirit, clearly defined by Mach in these words: to make abstractions meet perceptions, and to reconcile abstractions among themselves as economically as possible.

33

IS MATHEMATICS BY NATURE INCAPABLE
OF DESCRIBING REAL CHANGE?

by Jean Ullmo

LECTURER AT THE ECOLE POLYTECHNIQUE

MATHEMATICS experienced a stroke of misfortune unparalleled in the history of philosophic thought; it became the victim of a revolution which started out having nothing to do with mathematics and ended up dethroning mathematics from its former exalted station. From this derogation of mathematics some have sought to draw the gravest of consequences, casting scorn on intelligence and denying the value of human reason. One more step and thinkers would be calling upon man to surrender his ability to comprehend and yield instead to the forces of darkness. Let us try to shed some light on this "metaphysical drama," which, however, is not limited to the domain of metaphysics.

Until very recently the fundamental metaphysical presumption has been the distinction between Being and Becoming, and the acknowledged primacy of Being over Becoming. To have a temporal aspect and be subject to change seemed degrading. The higher reality, authentic Being, was conceived of as eternal, or better yet, as timeless. Time and Becoming were accidents which destroyed the perfection of Being. Plato's Ideas, Aristotle's Essences, Malebranche's and Newton's God, Kant's Noumena—all of these perfect beings escaped Becoming.

Mathematics benefited greatly from this metaphysical presumption, for it seemed to partake of that timelessness which was the symbol and the condition of perfection. When Descartes formulated what he considered the ideal rational explanation of the universe: *Causa sive ratio*—physical cause is nothing more than mathematical reason, "the relation between effect and cause is that between a conclusion and the principle from which it follows mathematically—he was thereby affirming that the progression of phenomena in time and the successive causalities of events were only illusions arising from man's limitations, his relative powerlessness. For to the inventor of analytic geometry, mathematical *ratio* was not prior to, but on the contrary, coexisted with the conclusion which it mathematically justified. Only the weakness of the human mind with its necessity for resorting to discursive thought introduces a before and an after into mathematical proofs. The moment the definition of the circle is given, so also are every one of its properties.[1] It is a failure in us if we need time to discover them. Likewise it is "our fault" if we discover these phenomena only in succession.

Through mathematics, therefore, we reencounter the timelessness of Being. For the great Cartesians of the 17th century—for Spinoza and Malebranche, for Newton, and even for Laplace—God is a mathematician. The phenomenon of time does not exist for Him. All epochs occur simultaneously from the standpoint of the Divine Intelligence, just as all mathematical properties coexist for this Intelligence, Who perceives them all at once.

In our lifetime we have witnessed a revolution in philosophy which has put an end to this primacy of timelessness. The great name of Bergson is associated with this revolution. To the supposed preeminence of motionless being and frozen essences, he opposed the profound reality of time in which the human adventure unfolds, and the irreducible value of "felt time" [*durée intime*], which is the be-all of man.

Fruitful as this revolution has been in many respects, it was however to have disastrous repercussions in mathematics. Since a whole philosophic tradition backed the claim that mathematics belonged to the domain of the timeless and was therefore the privileged instrument for exploring this domain, the conclusion was inescapable that mathematics was powerless in this newly discovered field of change [*devenir*] and life. Since the success of mathematical physics and of the prediction of natural phenomena could not be denied, the conclusion drawn was that things do not have *duration*, that we merely impose on them the semblance of time as we experience it [*durée vivante*]. Thus, though mathematics might adapt itself to suit the inferior domain of inanimate matter, it would remain absolutely powerless and inefficacious before the superior reality of life.

---

[1] Moreover, any one of these reciprocal properties can be made the definition and the steps of the proof taken in reverse order.

It was thus clear that the authority of mathematics was destroyed by this complete reversal in the viewpoint of philosophy. The harm would have been limited had mathematics been alone in suffering this loss of repute. But then again, everybody remained in agreement that mathematics was the grand creation of human reason, its most perfect manifestation and chosen instrument. To deprecate the tool was to humiliate the worker. Reason itself was proclaimed impotent by an entire school. After intuition there was a disposition to give preference to instinct, and then to many other things. The popularizers and zealots outdid each other in this drive toward obscurantism. Let us here attempt to ask them a preliminary question: Has their postulate been proved? Is reason so wretched a workman and mathematics so imperfect an instrument? Are they really absolutely incapable of understanding time?

It seems to us that such a conclusion rests on an outworn concept of mathematical analysis. Doubtless it comes from Descartes's analytic geometry. In this finite geometry a curve is completely given by its finite equation and the properties of the curve are indeed *inherent* in the equation and can be considered as given at the same time. There is no place here for a real succession.

But since the discovery of the infinitesimal calculus by Leibniz and Newton, the status of the problem has changed.[2] Today a mathematical representation of a phenomenon, or a theory in mathematical physics, is essentially knowledge of a system of differential or partial differential equations which govern the phenomenon; that is, its instantaneous change is known when one knows its present characteristics and the external conditions. *The mathematical representation describes the infinitesimal variation of the physical phenomenon.*

But, it will be said, this system of infinitesimal equations can be *integrated*; the equations of the phenomenon will then be known in finite form, and then the preceding conclusions will again become valid. Then it will be possible to think of the future yielded by the equations as inherent in the present, which has provided the means to set up these equations. This would not be a "real" future.

We believe that this is to misunderstand integration. Only in the exceptional case does the latter provide the laws governing the phenomenon in the form of finite equations. In principle, integration is essentially a method for obtaining by successive approximations the

[2] Cf. the following passage with the article by Brunet. (Note by F. LL.)

effect resulting from instantaneous variations, a progressive summation of elementary changes. It therefore follows the physical process of progression in time. The theorems on which the possibility of integration are based are related to developments in series. In principle these series furnish the means for calculating, step by step, the changes produced. But above all, these series—by their very definition—contain no more than these instantaneous variations which were furnished by experience linked with a progressive flow of time. They are nothing but a potent means for drawing conclusions progressively and successively, just as a future phenomenon stems progressively and successively from a present one. In infinite mathematics to deduce was to establish something. In infinitesimal analysis to integrate is to move forward.

Without going into the problem of liberty, which for Bergson is closely tied to that of duration, let us note that modern physics, with quantum mechanics, has been able to reserve a place in this progressive movement for the primary *free act* of observation (and the uncertainty principle which it entails), an act which constitutes an intervention of human consciousness in the forward motion of phenomena.

Thus the procedures of integration in mathematical physics reproduce—one might say step by step—the progressive and sequential unfolding which characterizes the world of observed phenomena and imposes on us the very concept of time. The course of thought merely precedes that of nature, the power of analysis abbreviates for us the duration of things without eliminating it.

One can even affirm that mathematics, rather than being an instrument for timeless synthesis, appears in its analytical aspect to be an instrument for *continuous exploration*, a path joining the near to the far, and that it would provide us with the notion of inevitable succession and consequently with the concept of time, if we had not already discovered this concept in our sense experience.

In order to render our conclusions more explicit let us take a closer look at the system of differential equations arising in classical mechanics, which have played such an important role in the age-old effort to understand the eternal world. The range and interpretation of these equations will become clear if we use the terminology of continuous groups of transformations,[3] considering them as representing

[3] A transformation transfers every point of a given space to another point. If these two points are infinitely close, the transformation is called infinitesimal; if

the infinitesimal transformation of a continuous group with one parameter[4]—which in fact they are. Integrating these equations gives us knowledge of the group itself. The best way to picture the progression and the result of integration is to imagine the infinitesimal transformation applied to an arbitrary initial state of the material system under study, and then applied to the resulting infinitely close state, and so on indefinitely, so as to propagate by this indefinite repetition of the infinitesimal transformation all the finite transformations of the group. This is tantamount to obtaining all the states through which the mechanical system passes, each state being associated with a fixed transformation and a corresponding value of the parameter.

We here see clearly that there is no identity and no inherency. The infinitesimal transformation *propagates* the group and *is not* the group. Similarly, a subsequent motion is not contained in the initial state; the two are not given simultaneously. The later motion *results* from the integration, i.e., from the repeated and successive application of the elementary change.

The group is built up by a continuous infinity of component transformations, each of these being associated with one value of the parameter. The group is propagated, as the future is born from the present, by a continuous modification. The variable parameter of the group, which regulates this propagation, may be identified with time; more generally, in relativistic mechanics, where time loses its privileged function, this parameter is redefined and constitutes the measure of change in the changing universe.

To put it a better way, for a mathematician outside of time, and knowing nothing of duration, the continuous increase in size of the parameter, corresponding to the progressive generation of the group, would furnish an absolutely clear intellectual model of the state of becoming.

Now that we have extracted this lesson from the intuitive representation of the generation of a group by infinitesimal transformations, we shall extract another from the method which is actually employed

---

not, it is called finite. A group is a set of transformations such that the transfer resulting from any two transformations of the set applied successively is equivalent to the transfer resulting from a third transformation which belongs to the set. Every transformation of a group is designated by a fixed value determined by one or more variable parameters.

[4] In the space representing the states of the material system being studied.

for integration. It obtains information for us about the entire group of transformations itself for all the values of the parameter. But this information in turn is to be analyzed in two stages. First we have to look for all the *invariants* of the group, that is, those things which do not change as the phenomenon undergoes change. (This is carried out by calculation of what are called first integrals.) But this element of identity is not the whole of the phenomenon. Knowledge of the invariants gives us the trajectories. To complete the description of the group we need a further integration to get the law describing these trajectories. Only the combination of these two aspects, the static trajectory and the kinetic descriptive law, gives a complete representation of the phenomenon.

Neither of the terms in this duality can be dispensed with. Mathematics conceived of in this way comprises not only identical or identifiable things, but also the dual notion of a change in which a part remains identical. Let us note, further, that among the first integrals (and the corresponding invariants) the most important is the one relating to energy; the conservation of energy under motion is just a special case of invariance. The preceding study shows the importance but also the limited significance to be attached to this invariance. Many philosophers seem to believe that the conservation of energy reveals the essential identity of the before and after in the unfolding of phenomena, the very symbol of the inherency that we translate as causality. We see that this is not so at all: Conservation of energy is but an invariance at the core of a transformation, a guide to the inner nature of a phenomenon and to its unfolding.

34

MATHEMATICS AND PHILOSOPHIC IDEALISM

by Paul Mouy

THEY say that when Pythagoras discovered the proof of the theorem which bears his name he made an offering of a hecatomb to Apollo. We do not in fact know the date, and the legendary personality of Pythagoras has no great bearing on the event; nevertheless, the occasion was worthy of note, for it marked the beginning of rational thought and the philosophy of idealism.

Doubtless many sources contributed to the origin of mathematics. They must be sought on earth and in the heavens: In the starry sky where the constellations presented the dual enigma of number and shape—of number and shape combined; on earth amidst the techniques of the surveyors and "earth measurers" (geometers), and of the bookkeepers who balanced incoming and outgoing goods and money for noble households and governments. There was a body of knowledge that sprang at the same time from techniques of measurement and "logistics," and from astronomical observations.

Indeed the feature which strikes us first about mathematics is that it is a body of *real* knowledge which deals with things and gets its teeth into matter, so to speak. It seems to be homogeneous with matter and to spring from it. A triangle is a reality. So is a number. They both have properties and a nature which seem to exist outside the mind, dictating to it their laws.

It is also likely that for a very long time, perhaps for several millennia, mathematics appeared to be an empirical art with elements of magic in it, like agriculture and medicine, a sort of efficacious witchcraft.

But the Greeks invented *mathematical proof*. Which Greeks? Perhaps Pythagoras, perhaps Thales. "The first man to consider the isosceles

triangle—whether he was called Thales or whatever—had an inspiration when he discovered that it was not necessary to stick to what he
saw in the figure . . . in order to conclude what its properties were but
that he had to construct this figure from his thoughts on the subject
and from his a priori concept. And to have certain a priori knowledge
of something he must attribute to it only what necessarily derived
from what he himself had posited according to his concept." This is
how Kant expressed it in his preface to the second edition of the
Critique of Pure Reason.

What in fact do we mean by proof? The first element is necessity.
Necessity, the Greek *Ananké*, is originally blind fate, which emanates
from circumstance and leads men to their destruction, which perfidi-
ously drove Oedipus to incest and patricide. A primitive idea. Thanks
to mathematical proof, this concept passes without change of name
from the exterior to the interior, from things to spirit, from the domain
of mysticism to the domain of reason. This was the concept which
exercised—against all reason—a compelling force upon man. It
becomes what man, by force of reason, compels himself to obey. It
constitutes an obligation on the part of the mind, an intellectual value,
value itself.

To prove also means to establish something a priori. The expression
"a priori" is a creation of the Scholastics, who used it to indicate the
direction of the movement whereby the mind goes from principles or
causes to conclusions or effects.

One cannot say that mathematical proof is always a priori in this
sense; it is not such when it proceeds analytically—as in algebra for
example, when one goes from the equation, which is the conclusion,
back to the root, which is its principle. But Kant gave to the expression
"a priori" the meaning universally given to it today: independently
of experience. In this sense every mathematical proof is a priori because it proceeds without calling upon experimental measurements,
such as evaluations of angles or lengths of sides. For a long time people
were satisfied with empirical evaluations of these quantities; among
certain Oriental peoples the value of π is equal to 3, and it is obvious
that this value was obtained by a rather crude measurement done with
ropes. But the Greeks were not satisfied with this approximation.
Through the same spirited thinking which led them to invent proof,
they conceived a rational approach, the method of exhaustion,
whereby the value of circumference compared to diameter could be
made as precise as one wished, without ever appealing to experience.

This was Archimedes' accomplishment; if he was unable to achieve it fully, if this required twenty additional centuries, it was because a totally rigorous solution required the use of algebra and serial expansion.[1] But the principle was won: you should be satisfied with what experience brings you; reason must reign as master.

However, all we have so far is a negative concept of a priori. As yet a priori signifies nothing more than non-experimental or rather non-empirical (for it is quite possible, and we believe it to be true, that the experimental includes the a priori as one of its ingredients.) But if a priori excludes everything that might come from *raw* experience—purely passive and to some degree naive—then we must believe that its value comes from elsewhere, therefore from the mind itself. In a positive sense, therefore, a priori means "by virtue of determinism or for a reason," as Hamelin puts it; better still, the a priori is "mind in its private dwelling place" [*l'esprit chez lui*]. Thus the a priori is the pure intelligible.

But what is the intelligible? What does it mean to understand something? We would be completely "at sea" with this question, as Leibniz phrased it, if we did not have precisely the aid of mathematics to give us our bearings and guide us. Furthermore, when we seek a model of intelligibility elsewhere than in mathematics—in logic for example—all we find is that the mind must not stop thinking what it thought to begin with, in order to be itself and remain faithful to itself. This is what is called *identity*. We would therefore arrive at the idea that intelligible means identical, or as the modern logicians say, that the highest principle of thought is the principle of *tautology*. But is it not obvious that this condemns the mind to sterility? Is it not especially obvious that this gives a very false picture of the workings of the mind in mathematics? If thinking meant identifying, all reasoning would be limited to keeping from losing, or losing as little as possible, of what one had included in his original premises. There would be an especial prohibition against adding anything whatever, an arbitrary enrichment being much more sinful than a wasteful loss. In short, a priori reasoning would only be a fortiori reasoning. This is, moreover, just what the Scholastics implicitly avowed in formulating the famous principle which constitutes the eighth and last of the rules to which the syllogism—i.e., reasoning about identities—must conform: *Pejorem sequitur semper conclusio partem*: the conclusion always follows the most

---

[1] On this subject see the article by Dubreil. (Note by F. LL.)

unfavorable part; it always takes on, and takes on totally, the loss in quantity or quality which tarnishes one of the premises. If one of the premises is particular and negative the conclusion can never be universal and affirmative. The first figure of the syllogism applies a universal rule to a particular or unique case or to any which is less universal or of lesser extension than the rule. The second figure allows itself only negative conclusions; it ends up by establishing nothing. The third figure authorizes only particular conclusions, which establish compatibility or incompatibility.

The procedure of mathematical reasoning, however, is entirely different. It recognizes only the universal, which it does not distinguish from the particular. "Triangle" is "every triangle." Moreover, in reality it always proceeds by way of "amplification," as the logicians call it; that is, it always ends in an increase of quantity. Logicians have racked their brains in a vain effort to discover a form of logical reasoning which really did "*amplify*." In the end they were forced to abandon their efforts to establish a true system of induction which permitted amplification. After all, that would be a contradiction in terms since logic makes intelligibility one with identity. But mathematicians solved the problem of amplification long ago. They proceed naturally from a specific universal to a generic universal; all work in mathematics has this orientation in common. Elementary geometry provides a good example. Observe how it starts with the triangle in order to conquer the polygon, starts with the square in order to "measure" the parallelogram and with the straight or broken line to evaluate the length of the circumference. Perhaps nothing is more characteristic in this regard than the famous proposition: *The sum of the interior angles of a convex polygon is equal to the number of sides of a polygon minus two multiplied by a straight angle.* This proposition is proved by starting with the case of the triangle, which the logicians see as merely a special case, like a logical corollary. And this without arguing in a circle.

One might ask, What is this "illogical" peculiarity of mathematical reasoning due to? To the fact, we believe, that the concepts which mathematics focuses on are considered solely as relations and never as classes, not even merely as substances whose properties we would like to predicate. The notion of relation is a very general and very intellectual notion. As has been seen by certain modern philosophers, who have attempted to give as broad and yet as precise an idea of the mind and its activity as possible, relation is intelligence in action.

To be sure, mathematics considers the problem of intelligence and intelligibility from a particular point of view and with a particular bias that simplifies the problem. The only relations that it considers are numerical, i.e., calculable by means of operations that reduce to addition. But from that beginning mathematics has reached so far and risen so high that one can say that it has given the human mind confidence in its powers and proved the conquering value of the *intellectus sibi permissus.*

In short, if idealism is defined as the philosophic doctrine according to which the universe is coextensive with representation, or put another way, if the intelligible is the foundation and the structure of reality, *idealism is mathematical.* It is mathematics which has brought mankind the revelation and consciousness of idealism.

But idealism is more than a philosophy of mathematics or even a theory of knowledge. It includes an ethics and a metaphysics. This ethics and metaphysics consist in faith in the spiritual.

The spiritual is the intellectual at its source, at the place where it springs forth from the mind. Now, it is here, at this initial point, that the mind finds God. The idea of God is an "innate" idea, the first innate idea, and it encompasses the evidence of its object, or rather it is one with it. The "ontological proof" in its true sense is the manifest presence of the infinite spirit in the individual spirit.

On the other hand, love is the soul of the intelligence. For intelligence is disinterestedness, generosity, a gift of the self. It is axiomatic that to understand is to love; in a sense it is to create or at least to recreate. Idealism attains its goal in God, and God reveals Himself as love.

This is the philosophy that we think mathematics embraces. To be sure, in formulating this we are not certain that we have the approval of mathematicians. They often think of themselves as realists or at least pragmatists because they confuse the attitude of fidelity to the mind with that of submission to things, and also because they shrink from posing the problem of mathematical truth; they prefer to consider themselves artisans rather than to set themselves up as secret priests of the intelligible world. But the philosopher has the right and the duty to strip them of their modesty and to lay bare the philosophy concealed within their science.

And now it behooves us to defend what we have been saying by the evidence of history.

Milhaud was quite right in naming as the founders of idealism the geometric philosophers of Greece, the Pythagoreans and Plato. The Pythagoreans clearly founded mathematical idealism with their well-known statement, *all things are number*. This statement has a twofold meaning: Number constitutes the intelligible structure of things, and the fundamental elements (we would call them categories) of mathematics are the elements of things. Now by virtue of the first meaning the Pythagorean principle establishes a rational foundation for mathematical physics, which by the way was also founded by the Pythagoreans through their work in astronomy and by their creation of *harmonic theory*, i.e., mathematical acoustics. With the second meaning the Pythagoreans assert the possibility of defining a structure of the mind that is a structure of things, made up of the concepts of finite and infinite, one and many, etc. Plato is of course the heir of Socratic conceptualism as well as of Pythagorean idealism. But it is rather clear that the latter outweighs the former in his philosophy by the emphasis he places on mathematics, amounting almost to collaboration, and by the place of importance he gives it in the education of the philosopher. The universe of mathematicians is a part of the intelligible world, an inferior part of course and a kind of prelude, the region of shadows and reflections. But Plato does feel that mathematics introduces the disciple to the intelligible world.

In the *Parmenides*, Plato will introduce a second philosophy, which he will expound in the *Sophist*. His aim here will be to provide a glimpse of conceptual dialectics, which has become for him the true theory of Ideas. He will bid farewell to the Eleatics and even commit the sacrilege of dialectically immolating the father of Eleatism; that is, he will declare that he is abandoning the logic of identity according to which *Being* exists and *Non-Being* does not, in favor of a metaphysics of "commingling," or in other words, of relation. This metaphysics appears to us inspired by mathematics.

Descartes ignores Plato, whom he considers the author of neo-academic skepticism. Thus one cannot speak, historically, of a Platonic tradition carried on by Cartesianism. But the fact, difficult to deny, that Descartes's philosophy is a metaphysics of *Mathesis universalis* makes him the unconscious descendant of Plato. Further, in the first half of the *Fifth Meditation* one can note expressions which come very close to Platonism, for these statements affirm the existence of an intelligible world composed of "true and unchanging natures." One even comes across an allusion, and at least a metaphorical adherence, as it were, to the Platonic theory of reminiscences. But that is not the essence of Cartesian idealism. The essence is *Cogito ergo sum*, which really establishes idealism by making every judgment about existence subordinate to and dependent on a pure and adequate intellectual consciousness. Beyond that, we can see by the *Regulae* that Descartes abandons the logic of identity, syllogistic logic, and that he bids it farewell as Plato bade farewell to Eleatism.

In short, Descartes's philosophy is a modernized Platonism, but still, like the ancient philosophy, based on mathematics. With *Cogito ergo sum* we have a Platonism that takes idealism as far as it can go. Indeed *Cogito ergo sum* is the very principle of idealism, for it means that reality is based on truth, and truth itself on intelligibility. Descartes certainly held this opinion, for he calls it "the first principle of philosophy."

Spinoza's and Malebranche's inspirations are quite different from Descartes's and from each other. But one point they clearly have in common; this is the conception they both form of extension, which in turn determines their conception of matter. To Spinoza extension is an "attribute of God," and he adds—perhaps with the intention of scandalizing his readers or in any event of arousing their attention— "God has extension." But for all this God does not become a material being, for the extension we are dealing with here is that of analytic geometry, an ideal extension, a system of ideas and relations. Likewise Malebranche holds that it is in God that we see bodies, because the extension he calls intelligible is God himself "to the extent that He can be partaken of by His creatures." To be sure, there is a considerable difference between the two systems: Malebranche distinguishes between intelligible and *perceptible* extension, the latter created and contingent. But the message of Descartes, the idealist message, has been understood and transmitted.

In a way Kant's philosophy marks a regression. Kant distinguishes a *Realgrund* or real foundation and an *Idealgrund* or intelligible foundation. Basically this is the substance of this criticism of the Cartesian ontological proof; reality is an absolute position and cannot be made to rest on ideas. In Kant's philosophy transcendental idealism is limited by empirical realism. This arises from the fact that the mathematics upon which he fashions his philosophy is no longer Cartesian but Newtonian. At the base of the latter lies an empiricism whose tradition goes back to Bacon. *Hypotheses non fingo* means basically that Newton rejects Cartesian idealism.

Newton abjures the Cartesian requirements of intelligibility. Universal gravity is an experiential fact; there is nothing about it to understand. To be sure, Newton starts by noting that gravitation might have a cause to be sought out, but he has *not yet* found it. His followers will be more categorical: In their eyes the problem does not exist; gravitation is an irreducible property of matter, and there is no need to search for anything beyond that. In the 19th century positivism will place a general prohibition on the search for causes; it will extend the interdiction of the Newtonians to the entire philosophy of science.

The interdiction has now been repudiated. The theory of general relativity is an explanation of gravity, an explanation that conforms to Cartesian requirements, for according to the general theory of relativity, gravitation is a property, not of bodies, but of space; it is a "curvature."[2] To be sure, we are far from the Cartesian vortex, but perhaps after all we are not so far from the intelligible extension of Malebranche; for this curved space has only an algebraic meaning; it is much less a state of being than a system of equations, a "world of numbers," as Brunschvicg puts it. In the history of gravitational theories we can note a constant, persistent tendency to separate weight from the body being weighed. We find it already in Galileo: When he maintains and proves—by curious reasoning—that bodies fall at equal rates (*in a vacuum*, we say, but Galileo had no vacuum pump; we should say *ignoring air resistance*); he means that weight resides in the gravitational field and not in the body. This is just how the Cartesians understood it. The theory of relativity identifies this field with space itself.

In short we seem to be witnessing a return to Cartesianism in contemporary science, and this return is likewise perceptible in the field of metaphysics. The very title of Husserl's *Cartesian Meditations* is evidence of this. We encounter again the profound import of *Cogito ergo sum*.

We may conclude with this: *perennis philosophia*, the philosophic tradition, has been absorbed into idealism to the extent that it was a philosophy of mathematics. It now tends more and more toward a form of idealism.

At this point we should describe the influence that philosophy has exercised upon mathematics. This influence is undeniable but complex and rather poorly known. We shall confine ourselves to a few suggestions.

[2] See the article by Thiry. (Note by F. LL.)

Philosophy has certainly had an effect upon mathematics by pro-
posing a certain ideal of intelligibility. This ideal has varied according
to the times. For example, Platonic idealism does not consist solely in
a certain requirement of intellectual clarity and discrimination; it is
fused with finalism. A Platonic idea is not only a relation capable of
being grasped by dialectical reasoning; it is also an ideal through
which the idea of the good manifests itself, and which exerts an
attraction upon the tangible world. A celebrated passage in the
Phaedo shows that in Plato's view an explanation of something consists
in demonstrating that a thing has to be the way it is by virtue of the
fact that it is good or preferable. Now in mathematics perfection
would be regularity and harmony. For example, Platonic idealism
translates itself, in its finalist form, by means of an ontological prerogative granted to the geometric solids presenting an interior
regularity—the regular polyhedra inscribable in a sphere and bounded
by surfaces that are regular polygons and in turn inscribable in a circle
and composed of equal and identically arranged elements. There are
five of these solids: the cube, and the regular tetrahedron, octahedron,
dodecahedron and icosahedron. In the *Timaeus* the five elements have
the shape of these five polyhedra, which was a problem even for Plato
since the tradition of physics provided him with only four elements
and his system required a place for the fifth. It is not by chance that
Euclid's *Elements* comes to a close with the study of the construction
of these five polyhedra.

In the 18th century Cartesian rationalism certainly exercised some
influence upon the creation of infinitesimal calculus. It produced a
kind of crisis in mathematical thought which was eased only when
Leibniz and Newton enrolled their new calculi in the ranks of *Cartesian Geometry*.

More precisely, it is even possible that certain theories that have
remained a part of science and have themselves submitted to positive
verification, first sprang up as part of philosophy. We are thinking
specifically of the wave theory of light, first enunciated by Huygens
in his *Traité de la lumière* (1690), which in our opinion is an offspring
of Cartesianism. Furthermore, this theory was almost immediately
picked up by Malebranche, who completed it and made it into a
theory of color.

Finally, it seems to us that it would be difficult not to recognize the
crucial influence of philosophical criticism upon the theory of relativity, which had so prompt a birth and so overwhelming a triumph.

Philosophy had been relativistic for a long time, in its own way, aspiring to logical coherence without the constraints of experimental applications, while science was oriented toward experimental applications, very often at a sacrifice of logical coherence. Then came the day when the shortcomings of the scientific system of explication broke out into the open. It was then the job of science to find the new mathematical forms applicable to experience. But perhaps scientists would not have shown any concern in this regard if they had not been subjected to the suggestions of philosophical criticism.

So the relations between philosophy and mathematics are reciprocal. They exchange services and influence each other by turns. For the good of the human spirit we must hope that they collaborate without misjudging one another.

35

MATHEMATICS AND MARXISM

by Paul Labérenne

HAVING had occasion, some twenty years ago, to speak of Marxism and mathematics with a well-known mathematician and philosopher of that period, I saw him smile. "Ah, yes," he told me, "dialectics! — times — gives +. I once had a Socialist friend, a brilliant professor of literature, who thought he could convert me to the beauties of Marxism with this algebraic example, which he thought impressive."

Dialectical materialism is somewhat better known in our time. However, I am not sure that certain of its adversaries and even some of its defenders do not still sometimes reduce it—particularly where mathematics is involved—to examples almost as simplistic as the one we have just cited.

Although the founders of dialectical materialism were not mathematicians by training, the elements in their work pertaining to the exact sciences already went beyond those "rigid trichotomies" that Marx himself ridiculed in a letter to Engels (January 8, 1868) concerning the works of Stein.[1] Engel's *Anti-Dühring* contains among other things some very penetrating remarks on differential and integral calculus, and we know that toward the end of his life Marx made a very deep study of higher mathematics with the twofold objective of putting into algebraic form the economic laws he enunciated in *Das Kapital* and of studying some of the modes of argument of mathematical analysis from the point of view of dialectics.[2]

---

[1] In his *A propos de la dialectique* Lenin, too, regrets that the identity of contradictory principles is too often considered by Marxists simply "as a collection of examples," with the result that the autodynamics of this law is not made evident. (Lenin, *Œuvres complètes*, tome XIII, édition française; E.S.I., 1928, pp. 324–325.)

[2] Marx has left around 900 pages of mathematical manuscripts which have

Indeed there is nothing surprising in this interest of Marx and Engels in mathematics if one reflects upon what dialectical materialism really is. This conception of the world, which we can consider the most fully realized system of modern rationalism, encompasses all aspects of nature and thought in their own dynamics and in their mutual relationships. While the majority of philosophies study the givens statically and try to put them into a formal and narrow framework (this is notably the case for the positivism of Auguste Comte), Marxism is a philosophy of movement. Enlisting the celebrated Hegelian triad into its own service, Marxism studies phenomena and concepts in their development, affirmations and contradictions; each contradiction is in the end "lifted up," and leads to a synthesis richer than the original thesis. But while Hegel gives ideas primacy over nature, Marx, according to his well-known phrase, "sets dialectic back on its feet" by giving nature the primary role, without however forgetting the importance of the interaction between nature and thought. The evolution of inorganic matter, of the organic world, of consciousness and thought, the history of man himself, thus form part of one and the same whole: Starting with material and economic infrastructures, political and intellectual superstructures arise, among others science; this entire development occurs, not in a uniform or regular pattern, but with sharp interruptions in the march of progress, with sudden changes from quantity to quality, with crises, and with revolutions.

Now, despite the extremely abstract aspect that mathematics presents at the present time, it provides some remarkable illustrations of the correctness of the Marxist thesis.

Thanks to dialectical materialism it is in fact possible to discover beneath this abstraction the economic and social factors underlying the historical development of this science, by laying bare its links with technology, and the relations between its evolution and human evolution in general, including also the various class struggles.

Thanks again to Marxism it is possible to acquire a better understanding of the essential features of what seems to be the characteristic behavior of mathematical research and to discover the real rhythm of the great discoveries in mathematical thought.

Finally, if we now apply Marxism not to the past but to the present

been translated into Russian by the Marx-Engels Institute of Moscow. They consist principally of résumés of the then celebrated texts on analysis by Lacroix and Boucharlat, with certain more personal notes on the operations and symbols of differential calculus.

and the future it will permit us to analyze the nature of the difficulties encountered today by mathematics and to sketch the broad outlines of a solution.

All these various aspects of dialectical materialism as applied to mathematics are closely linked to each other; to remain faithful to Marxist thought we would have to study them practically simultaneously. For the sake of clarity, however, we shall separate them, and since other articles in this collection, in particular Chapelon's, deal with the first point, we will treat the latter two in more detail.

As we have just said, differential and integral calculus received Engels's attention in his *Anti-Dühring*. He wrote:[3]

> The mathematics of variable quantities, the most important part of which is the infinitesimal calculus, is essentially nothing but the application of dialectics to mathematical questions.

And he laid stress, in particular, on the "negation" of x and y which the introduction of the quantities dx and dy leads to, as well as on the negation of the negation to which integration, which gives us back the finite quantities, corresponds. In addition, he emphasized the solution of a "problem which ordinary geometry and algebra had already broken their teeth on."[4]

Actually Engels could have cited the entire history of the concept of number.

We shall limit ourselves to the great epochs. We recall that the first serious difficulties arose in classical antiquity with the appearance of irrational numbers.[5] Up to then Greek mathematicians were familiar only with integers and fractions. Adhering to the simplistic theories developed by the Pythagoreans, they believed that "the point was the unit of position" and that every straight or curved line segment should be thought of as a kind of necklace made up of a finite—though perhaps very great—number of tiny particles of equal size. It followed from this that all line segments were commensurable with one another, and that measuring them could lead only to what we now call rational numbers.

---

[3] Engels, *Philosophie, Economie politique, Socialisme* (contre Eugène Dühring), translated by Laskine, Giard éd. 1911, pp. 167–168.

[4] Engels, *loc. cit.*, p. 172.

[5] It will be helpful to compare this entire discussion of irrational, negative and complex numbers with the article by Fréchet, "The Natural number and its Generalizations," Part I, Book Two, pp. 70–80. (Note by F. LL.)

The discovery of the theorem about the square of the hypotenuse, which is attributed to Pythagoras personally, dealt a mortal blow to this conception when the inescapable conclusion was drawn from it that the diagonal of a square is not commensurable with its side. They say that the terrified Pythagoreans at first took an oath to guard the secret of the discovery of these irrationals, for fear that the Supreme Architect would seek vengeance if they revealed his imperfection.[6] *Alogon*, the inexpressible, was the name given to the irrational numbers.

Even so the secret spread. Greek mathematicians finally accepted these mathematical entities whose existence seemed to negate the very concept of number as they had entertained it up to then. They thus arrived at the most general concept of *arithmetic* number, a concept infinitely richer than that of *rational* number. But we had to wait more than 2000 years for the German Dedekind to appear with his theory of cuts and give us a theoretical justification of the legitimacy of the operations already performed by the Greeks upon irrational numbers.

Similarly, negative numbers were introduced in contravention of all the philosophical principles the Greeks held dear. The Hindus and the Arabs, who were the first to use them, were not embarrassed by logical scruples. But when European scholars, at the beginning of the Renaissance, set about retrieving the classical heritage, they were somewhat shaken when it came to integrating it with Arabic algebra. For example, the German Stifel, who seems to have been the first to clearly present the idea that negative numbers are less than zero, as late as 1553 called them "absurd numbers," as opposed to positive numbers, which he called "true numbers." Descartes himself, almost a century later, will style them "false numbers." However, this second crisis ends with the absorption of the new "absurdity" into the concept of number, which is thereby enriched once again.

What are we to say then of the introduction of complex numbers, which Cardan will first consider when studying the roots of third-degree equations? Some years later Bombelli establishes the first rules for operating with complex numbers, in connection with the equation $x^3 = 15x + 4$, while confessing that for a long time they had seemed a "senseless idea" to him. Long after Gauss had found a geometric interpretation for these new numbers mathematicians continued to think that they were really absurdities whose usage was justified only pragmatically. Are they not still commonly known as *imaginary*?

[6] On this point and for the whole history of number in general, see the interesting book *Number* by Tobias Dantzig (The Macmillan Co., 4th ed., 1965).

The same outcries and the same scares accompany the invention of differential and integral calculus, which, as we have just said, Marx and Engels saw as one of the most beautiful examples of dialectics. Mathematicians allow finite differences to collapse to zero, yet they continue to divide one by the other by means of the ordinary rules. They calculate the lengths of arcs of curves by thinking of them as the sum of an infinitely large number of infinitely small straight line segments. They calculate a plane area by interpreting it as a sum of an infinitely large number of infinitely narrow rectangles. These first discoveries are very exciting, and stimulate a veritable frenzy of more or less disorganized mathematical creation in the 18th century. Some, like Bishop Berkeley, have misgivings. With respect to Newton's theory of "fluxions," Berkeley asks whether mathematicians "do not take things upon trust, and believe things," and whether they have not "their mysteries, and what is more, their repugnances and contradictions."

But the mathematicians have once again thrown off the yoke of the old Aristotelian logic, and their joy in creativity is greater than the fears of philosopher-bishops. "Go forward, faith will follow!" said d'Alembert, and in fact discovery succeeds discovery and errors are rare. When this heady creative period is over and the mathematical discoveries of the preceding centuries are filtered through 19th-century criticism, the balance sheet will be impressive. The "mathematics of variable magnitudes," in Engels's words, has made a prodigious leap beyond traditional logic and the concept of number has become enriched by the notion of infinitely small quantities and the possibility of summing them.[7]

The most recent of those fertile contradictions which confront the science of numbers at each new stage are to be found in set theory.[8] Here mathematicians make a renewed effort to grapple with the "unthinkable" and Cantor attempts to classify infinite quantities. We know what impassioned discussion these studies provoked in their day. Moreover these discussions are not completely at an end, for there still remain several contradictions which one can avoid only by refusing to consider certain mathematical concepts such as the set of all sets or

[7] In fact, paralleling the enrichment of the concept of number is the enrichment of the concept of operations. The study above is at best a quick sketch and requires amplification.

[8] In connection with the passage which follows see the articles by Eyraud, Denjoy and Buhl. (Note by F. LL.)

the set of all ordinal numbers. Furthermore, there are still many difficulties in this field remaining to be solved. The subject is full of traps and perhaps more than in any other domain the mathematician must strive to think dialectically. Engels had already said, "Infinity *is* a contradiction, and is full of contradictions. Eliminating the contradiction would be the end of infinity."[9] Therefore for anyone working in set theory the problem is not to "suppress" the contradiction but to "lift it up" by integrating it and by transcending the two terms in a new enrichment of mathematical concepts.

Constructed initially upon an extremely abstract foundation, for a brief time it seemed that set theory could not survive the contradictions to which it led. But with the first applications of set theory to other branches of mathematics, notably topology, its essential principles finally won acceptance. The great majority of contemporary mathematicians would now subscribe to the opinion expressed by Hadamard in the course of the celebrated debate of 1905 in which Borel, Lebesgue and Baire likewise took part. "I believe that the debate is fundamentally the same as that which arose between Riemann and his predecessors concerning the very notion of function. . . . It seems to me that the truly essential progress of mathematics, starting with the invention of infinitesimal calculus itself, has consisted in the annexation of successive concepts which in varying ways for the Greeks, for Renaissance geometers and for Riemann's predecessors, were 'outside of mathematics' because it was impossible to describe them."[10]

Limited space prevents me from going into other examples. I shall have to be content with pointing out, in geometry, the difficulties that mathematicians experienced in denying Euclid's postulate and in conceiving a richer geometry of which Euclid's would be only a special case. As early as the 18th century, Saccheri and Lambert do in fact discover the principles of non-Euclidean geometries, but they reject the implications of their discoveries for puerile reasons (properties repugnant to the "nature of the straight line") or because of grossly incorrect reasoning—which is surprising, coming from them. John Bolyai himself, after laying the foundations for his "absolute geometry,"

[9] Engels, *loc. cit.*, p. 46.

[10] The letters exchanged during this debate have been reproduced in Borel's *Leçons sur la théorie des fonctions*; the letter of Hadamard alluded to in the text is on page 157 of the 3rd edition (Gauthier-Villars, Paris, 1928). If more space had been available we would also have devoted equal attention to Hadamard's discerning remark concerning Riemann's generalization of the notion of function.

seeks to prove the famous postulate by starting with the supposed contradictions in the relationships among the distances of five
points in space. It takes Lobachevsky almost 30 years to abandon the
name "imaginary geometry" and call his collected works "pan-
geometry." However, we know the end that relativity had in store for
"imaginary" or "absurd" geometries.

But not only does dialectical materialism explain the progress of
mathematical thought leading to discovery by placing it within a
more general scheme; it also permits us to understand where these
sudden enrichments come from. Its analysis, founded on the interdependence of the sciences and on the action and reaction between
mind and nature, eliminates the "gratuitous" and purely "formal"
character which is too often attributed to certain discoveries. Despite
its abstractions, mathematics is steeped in reality (the word reality
being used in its ordinary sense).

"*Pure* mathematics," as Engels already said,[11] "is, after the event,
applied to the universe although it derives from the universe and
represents but a portion of the latter's actual forms and combinations,
and, moreover, *it is for this reason alone*, that mathematics is applicable
to the universe."

Other articles in this collection give a more detailed exposition of
the concrete origins of mathematics and the determining influences
brought to bear on mathematics by the external world throughout the
various epochs of history (influences often not very clear to mathematicians themselves).

What I would like to emphasize here is the fact that the knowledge
of the technical and social foundations of the historical development
of mathematics, together with a deepening understanding of the
dialectic movement of thought, lend to this subject a completely new
impetus and power. On the one hand, an awareness of this dialectic
movement is in fact an additional aid to the scholar in his research.
On the other hand, a presentation of its links with nature and with
human activity leads one to view the progress of mathematics from a
more general point of view and to fit this science into a vast program
in which it will develop in direct and close relationship with the needs
of other sciences and of society, never, however, sacrificing the construction of great and necessary theoretical foundations solely for
immediately useful applications.

[11] Engels, *loc. cit.*, p. 25.

The complete realization of such a program implies a total transformation of society, and it is understandable that this was first possible to envision in the U.S.S.R. The books and articles by Professor Colman of the Moscow Institute of Mathematics and Mechanics provide us with an important record of the whole movement of ideas in this direction on the part of Soviet scientists.[12]

Colman believes that there is at present a crisis in mathematics which can be resolved if mathematicians direct their work more along the following lines:

1. Research on a synthesis between the continuous and the discontinuous; this research is more and more necessary because of recent progress in quantum theory. Efforts already made in this direction in set theory appear in fact inadequate, inasmuch as they remain too abstract and are conceived from too idealistic a point of view (using the word in Marx and Engels's sense). Colman finds that in the final analysis the general regression of bourgeois ideology is the cause of the idealism of numerous contemporary mathematicians, for whom the subject seems no more than some kind of game. This ideology, he says, was progressive at the beginning of the last century, but became more and more reactionary through fear of the proletariat, and now maintains an ironic attitude toward science and progress.

2. Research on a synthesis between the analytical or "static" aspect of laws and their "dynamic" or statistical aspect. The calculus of probabilities still seems too much separated from the rest of mathematics although it is playing a larger and larger role in all applications of mathematics.

3. A return to the particular, which has been unduly neglected in favor of a more and more undifferentiated generality. In vector analysis, Colman sees the first instance of such a return and its fruitful consequences, and he would like to see analogous mathematical methods appear in other fields.

4. Likewise a return to the historical origins of the various concepts, which will keep us from succumbing to the opinion that discoveries

[12] Let us cite one article among others, "The present crisis in mathematics and an outline for reconstructing it," published in *Science at the Cross Roads* (Kniga, London, 1931), a collection of papers presented by the delegates of the U.S.S.R. to the International Congress of the History of Science and Technology, held at London in 1931. We also wish to cite the book *Predmet i Metod Sovremennoi Matematiki* (Aim and Methods of Modern Mathematics), published in Moscow in 1936.

are to some degree "gratuitous," an opinion that is injurious to the methodical development of the science.

5. An effort to bridge the gap between theory and practice. This gap tends to widen as a consequence of extreme specialization and also because of the intellectual's "bourgeois" scorn for applications.

6. A thorough dialectical investigation of the problem of mathematical foundations, which implies a struggle against Weyl's and Brouwer's intuitionism with its mystic tendencies and the sterile skepticism of Russell's formalism.

In his writings Colman then sketches a more specific program which recalls the programs of Erlangen and Paris. Among other things he proposes:

From the historical point of view—a detailed study of the development of contemporary mathematics in our epoch, the epoch of imperialism, in order to gain a better understanding of the causes of the present deficiencies and weaknesses of contemporary mathematics;

From the point of view of the synthesis between continuity and discontinuity—a more thorough investigation into topology, which might link set theory to other areas of research, as well as a return to certain neglected portions of mathematics (theory of equations with finite differences, theory of divergent series, etc.);

For the purpose of strengthening the bonds between theory and practice—a great increase in the number of mathematical institutes applied to industry, agriculture, topography, statistics, etc., where mathematicians would constantly have to solve concrete problems requiring the simultaneous application of various branches of the science, (this would also combat the tendency toward specialization).

Many of the objectives named by Colman correspond precisely to the directions in which Soviet mathematicians had been working in the years preceding the Second World War. We know, in particular, the remarkable progress they had made in topology and in the theory of numbers (Goldbach's theorems; the transcendental nature of a to the power b, where a and b are two algebraic numbers and b is irrational; topology of the theory of groups, topology of set theory, topology of hyperspaces, and calculus of variations).

Czarist mathematics already occupied a position of honor in world science; Soviet mathematics occupies a place of the very highest order. Two or three other foreign schools can be put on a level with the Soviet school; none of them at present can be considered superior to it.

Without doubt, the very fact that the subject is no longer reserved for certain privileged beings and that, thanks to the establishment of socialism, the number of investigators has multiplied and their recruitment has been considerably improved, has a great deal to do with this. To be sure, the fact that all Soviet science is absorbed in the task of building a new world also explains in part the creative enthusiasm of these young scholars. By their daring they remind us of the young polytechnicians whom Monge trained at the time that the Revolution and the Empire were giving the old feudal Europe a vigorous jolt.

But besides these factors, it would be a mistake to overlook the direct influence of training in dialectical materialism. To be sure, as Engels said, "Men thought dialectically long before they knew what dialectics was, just as they spoke prose long before they knew the term."[13] And what is true of men in general is perhaps even truer of mathematicians, as the examples we have given in this very article tend to prove. However, programs like Colman's and the frequency with which efforts are directed systematically towards new syntheses show a conscious desire on the part of young Soviet mathematicians to utilize dialectical materialism as a tool in their research. Moreover, Kolmogorov, Alexandrov and Schnirelman, to mention no others, have frequently acknowledged, in the review *Front Science-Technique*, the advantages they have procured from their knowledge of Marxism.

Thus, Marxism is not satisfied with giving us an explanation of the historic evolution of mathematics as a function of the technical, economic and social conditions on which it depends; it also permits us to analyze the very mechanism of the progress of scientific thought through successive crises and syntheses and to give direction to our research. In this it is more than one of those philosophic conceptions which seem especially good at justifying and classifying what already exists; it contains within itself a creative dynamism, the fruits of which we have already begun to see in the Soviet state. As far as mathematics is concerned, Marxism sometimes is doubtless still close to those simplistic schemes we spoke about earlier. But the study of science in the light of Marxism is still in its beginnings, especially in countries other than the U.S.S.R. There is still much to be done in investigating the fine points of the dialectic approach and in correctly applying the

---

[13] Engels, *loc. cit.*, p. 179.

new intellectual instrument to the directing of research. But the results already obtained, especially by Soviet scientists, already show the power of this new instrument. Thanks to Marxism, an era of modern, dynamic and constructive rationalism will succeed the era of static rationalism with its rigid, logical or logistical framework. The social revolution today taking place in the midst of all the tragic upheavals of the contemporary world mirrors a mental revolution which amounts to a veritable breakthrough in man's knowledge of the way his thought progresses. The two revolutions have the relation to each other of two aspects of a single phenomenon. At the moment we are more impressed by the former, but we are sure that the future will soon show that the second one is just as profound and just as rich in fruitful consequences.[14]

[14] The reader will find supplementary details on certain points of this article in two articles we have published in *A la Lumière du Marxisme*, Volume I, under the titles "Les mathématiques et la technique" and "Le Matérialisme dialectique et les sciences." (Éditions Sociales Internationales, Paris, 1936.)

Book Three

TRUTH AND REALITY

MATHEMATICS AND THE
NATURAL SCIENCES

If we had intended the relative lengths of the various subdivisions in this collection to reflect accurately the relationship between mathematics and civilization, we would certainly have had to develop much more than we have the two sections dealing with the relations of mathematics with science and technology. We have avoided doing so, for it seemed to us that this is an aspect of the role of mathematics fully accepted—if not well known—by the public at large. Leaving aside some topics to which we shall return in a sequel,[1] we have been content to emphasize the exchanges which have taken place recently between mathematics and the so-called exact sciences.

The three studies that we present on this subject are preceded by a more general article which will form a natural transition from the section on philosophy.

A good classification of the sciences is more than a table of contents; it presupposes a certain notion, first of the origin of the various scientific ideas, and then of the influence they can and ought to exert upon each other. By virtue of this, the classification constitutes a position both philosophic and strictly speaking scientific which serves as the unseen underpinning for many an effort, whether

---

[1] In particular, topics dealing with the relation of mathematics to biology, psychology, political economy, history, pseudo-sciences, the classic exact sciences, the concept of reality, etc., which we shall assemble in a second series [never published].

sterile or crowned with success. How does one measure the influence of a more or less vaguely perceived general thesis—a mental atmosphere, as it were—upon the discoveries being elaborated within one's soul? It is never a waste of time to try to establish the position of an object of value among a great many other things. The position of mathematics is by general consent particularly important, for mathematics (with or without logic) provides the starting point for the other sciences.

There is no need for us to say anything in these pages about Raymond Queneau's rare qualities as a writer. His curiosity ranges over a wide area of interests including mathematics, as several of his works have made evident. The classification which he undertakes to justify in his article upsets, to a certain extent, the traditional one of Auguste Comte and is to be compared with the positions recently maintained in the Soviet Union. This procedure, by viewing mathematics as the destination rather than as the point of departure of the sciences, has the great advantage of offering opportunities for renewal, whose premises are perhaps not impossible to discern.

The history of science shows that the great advances in physics are generally due to the application of a new and appropriate set of mathematical tools to a new experimental subject. The forms that these mathematical tools take, and the methods for applying them, are generated and determined by the particular properties and kinds of resistance shown by the subject which they are called upon to master. Most often the physicists place an order with the mathematicians, specifying their needs; this is one of the best opportunities for an enrichment of mathematics. Sometimes it also happens that in some back room of the mathematics storehouse, the physicists discover a somewhat neglected and unappreciated theory or algorithm, which then wins high favor overnight. Think, for example, of the Riemann-Christoffel tensor conceived a century ago, and how we had to shake off the dust which covered it so that we might assign to it the glorious responsibility of representing the curvature of the universe in the general theory of relativity.

During this first half of the 20th century physics has been shaken by resounding explosions which have profoundly and permanently modified our conception of the material world. To conquer these unknown realms it has been necessary for the scientists to call upon a whole new arsenal of arms. In the new panoply of contemporary physicists one finds such algorithms as matrices, vectors, tensors, operators, etc.; theories such as non-Euclidean geometries, Riemannian spaces, Hilbert spaces; and even complete bodies of doctrine such as the theory of groups and the calculus of probabilities. The development of such a flourishing commerce between mathematics and physics demanded a special audience.

We have asked for three articles focusing on this subject. The first of these presents a wide survey, while the two others establish connections, one with traditional mathematics and the other with the future. Thus we establish simultaneously the continuity and the revolutionary character of the spectacle we have the privilege of witnessing.

Louis de Broglie needs no introduction. The creation of wave mechanics—which won him the Nobel prize—by clarifying quantum theory has resuscitated modern science and through it our view of the universe. In his clear and easy style, which perpetuates the grand tradition of French scientific writers, Louis de Broglie examines from a very general point of view the role of mathematics in the development of contemporary physics. His incomparable mastery of the subject permits him to cut broad avenues through this heavily wooded forest. Modern physics is reduced to its three essential aspects—particle, relativity, and quantum physics—which correspond to two essential aspects of mathematics—continuity and discontinuity—and to their synthesis. De Broglie reviews the mathematical tools that are particularly adapted to each of these areas and in doing so provides us with a very instructive, rapid synthesis.

Maurice Janet is one of our leading experts on partial differential equations and on the calculus of variations. He has devoted two sections of the Mémorial des Sciences Mathématiques *to a related theory, the theory of integral equations. He there presents a masterly study of the case of symmetrical kernels (to which he alludes in his article). We have requested him to present this modern extension of the classic theory of differential equations because this theory, through the medium of Hilbert spaces, is curiously joined to matrix analysis, and matrices have been shown to be a valuable tool in de Broglie's wave mechanics.*

As for the calculus of probabilities, it is first of all a keener scalpel placed by mathematics at the service of classical physics. The success of the kinetic theory of gases in the 19th century attests to the value of this tool created two centuries earlier. But the new probabilistic conception of physics goes much farther, and it urges us to modify our way of thinking profoundly. It is no longer a matter of using the calculus of probabilities as a wonderful makeshift which furnishes sufficiently good approximations and results for practical purposes, giving no thought at all to the foundations of our natural laws. Now, on the contrary, it is these foundations themselves whose validity is being contested, and the principle of determinism itself which is under attack. The entire structure of the philosophy of science—and because of this, of all of philosophy—is being called into question. Is it possible to sail clear of the headland of an outworn faith in a petrified mechanism, while avoiding the reef of a retreat toward some mystical irrationalism of the past which is even less desirable?

To clarify this discussion and guide it to the right conclusion required a subtlety and objectivity that we knew we could find in Théo Kahan. Receptive to all aspects of culture and thought, Kahan has concentrated his research in the fields of nuclear physics and ultra-short waves. His work has familiarized him with both theoretical and experimental physics. His compact and brilliant study will permit us to extract the leading ideas in this difficult problem which is certain to furnish some of the specific features of the scientific humanism of the future.

F. LL.

36

THE PLACE OF MATHEMATICS IN THE CLASSIFICATION OF THE SCIENCES

by Raymond Queneau

In its relations with mathematics every science goes through the following four phases (four up to now, perhaps five tomorrow): *empirical* when facts are recorded, *experimental* when they are measured, *analytical* when they are calculated, and finally *axiomatic* when they are deduced (from premises of a meta-(science) or a system of logic relevant at the time). During the first period, mathematics plays only a lowly role, with arithmetic at the very most entering into it. In physics fluids are enumerated; in chemistry, elements; in biology, species; in psychology, spiritual and mental faculties. Next mathematics enters at the level of geometry and algebra, and go from enumerations to formulas. Mechanics and astronomy now come into being; in physics, we get Mariotte's law; in chemistry, we have weighing and Lavoisier; in biology, Malthus; in psychology, Fechner. Only physics has entered the third phase in its entirety; having finally discovered with precision what it is that must be calculated, it borrows from "reality" (formulated at previous stages) a certain number of concepts to which to apply the analytic method. Let us understand by this that first a Leibniz-Euler analysis is employed, and then a Newton-vector analysis. Formulas (algebraic) are supplanted by differential equations. For scholars at the end of the 19th century this is the "ideal" stage. The other sciences remain far behind in this respect. Only very small areas are studied analytically: in biology, the theory of the struggle for life; in sociology, econometrics. These two examples show, moreover, that there is nothing incompatible between the analytic method and the life sciences. A partial explanation of its slow rate of application to them may be that the analytic method does not seem to involve the solution of any mathematical problem; mathematicians are not interested in fields which offer no prospect of mathematical discoveries. If Volterra developed the theory of the struggle for life, it is because it led to integral-differential equations worthy of interest. As for econometrics, it was natural to use number in the domain of economics. However, if this science still shows relatively little development, it is because mathematicians have not found it to their taste. "For mathematics to dominate a science, not only must mathematical reasoning be necessary to it, but this reasoning must also be interesting to mathematicians." (G. Evans, *Stabilité et Dynamique de la Production dans l'Economie Politique*, Mémorial, Vol. VLI, p. 2).

The ever-widening use of the calculus of probabilities is now somewhat distorting the above picture, which was correct up through the early years of the 20th century. The biological and sociological sciences believe that they have found in this method the suitable instrument for their mathematization and, compared to analysis, more flexible and better adapted to the life sciences. To tell the truth this is an illusion. For example the applications of statistics to psychology, interesting and cogent as they are, have perhaps done less to elevate this almost nonexistent (scientifically speaking) discipline to the rank of an established, valid science than has a theory like behaviorism. There is no doubt that biology and sociology will have to become mathematized if they are to be true sciences. Will they do so aided only by statistics or is their only way to be poured into the same mold that already produced physical chemistry? However, this is not the question we should be asking; we shall see why below.

If we come back to physical chemistry we see that it is at this moment seeking to axiomatize itself and in this way to render itself indistinguishable from mathematics, which did not reach this stage until nearly the end of the 19th century. But in this very matter we find that mathematics no longer occupies a superior position; its relationship with logic appears to be one of dependence as a rule, and in any case is rather ambiguous. It is incontestable that logic became a science only by being transformed into logistics, i.e., by passing from the empirical and Aristotelian stage of enumeration to the experimental stage of formulas, i.e., by using geometric-algebraic methods (and one cannot see how it could go further, and yet there must be something further). Use of these methods, however, makes it dependent on a science over which it legitimately rules since it must establish the principles of that science as of every other science.[1]

But what is logic if not a science of "thought" and an art of "thought"? As a science of "thought" it must be no stranger to psychology; we leave consideration of logic as an art for later. Thus, when science has advanced as far as it can, we find ourselves back at psychology. This can be viewed from another angle. Let us consider the relative positions of sciences on a descending scale; we can say— as has been said many times—that if the ideal of physical chemistry was to become mathematized, it was the wish of mathematics to absorb physical-chemistry; the axiomatization of the latter would be the final culmination of this wish. Similarly, physical chemistry wants to absorb biology and biology wants to be incorporated into physical chemistry. The discovery of ultra-filterable viruses thus satisfies those who hold this double point of view. Finally, there is no doubt that the miscellaneous anthropological sciences, particularly psychology, aim at being reduced to biology. There is therefore a double movement all along the scale; this movement has been perfectly described by Meyerson and finally ends where?—in logic. If we admit that "logistics is the axiomatics of thought itself" (Piaget, *Classes, Relations et Nombres*, p. 5) we suddenly find ourselves facing a science, psychology, that we had at first considered the most backward of all, scarcely at the experimental stage; now, from a certain point of view, it would seem to have suddenly attained the highest level, the axiomatic, and this by virtue of its very nature.

The essence of logic is twofold; if we view it as dogma, we arrange all the sciences in a curve which forms a closed circle. Having reached the summit, we fall back on human contingency. And what would a mathematized and then axiomatized psychology be if not perhaps logic? But logic is also an art, and axiomatization a game. The ideal concept of science, as scientists of the early 20th century conceived it, was not a body of knowledge but rules and methods: notions (un-definable), axioms and the method of using them—in short, a system of conventions. But what is that if not a game, differing in no respect from bridge or chess? Before examining this aspect of science further, we must stop at this point. Is science knowledge, does it serve the process of knowing? And since we are dealing with mathematics in this article, what does one know in mathematics? Precisely nothing. And

---

[1] See the article by Marcel Boll and Jacques Reinhart, Part III, Book Two, pp. 26–41. (Note by F. LL.)

there is nothing to know. We know no more about point, number, group, set, function than we do about the electron, life and human behaviour. We know no more about the world of functions and differential equations than we "know" concrete, mundane, daily reality. All we know is a method that is accepted (agreed upon) by the community of scholars as true, a method which *also* has the advantage of joining up with manufacturing techniques.

But this method is also a game, quite literally a game of the mind.

Thus all of science, in its most finished form, will show itself to be a technique and a game, i.e., just what the "other" human activity, art, shows itself to be. Art, too, balances between these two poles: "art" in the strict sense, craftsmanship, and the gift of "inspiration" as well as that of "execution." There are "arts" which are "useful" techniques, others which are "less" so; just as chemistry is "useful," for example, in so far as it aids in the preparation of medicines, whereas conic theory was not at all "useful" as long as it did not serve astronomy and the nautical art. In and of itself science taken as knowledge finds itself occupying the same position as works of art which are also occasionally made to pass for a kind of knowledge. Both are *fictitious*, whereas a game is *factual*, and technology *effective*. Again, giving art its ambiguous meaning, we can say that science oscillates between art and game, while art oscillates between game and science. But then what is knowledge? And what is a game? And finally, what is technology? And wherefore these activities? It is hardly possible to perceive a meaning in these questions as long as psychology is not mathematized. On the other hand, it would be desirable to see esthetics place itself on the same level as logic, i.e., *establish* its own validity.

In other respects it is not evident that the mathematization of science operates in the fashion we have just described, i.e., by employing "applied" mathematical structures and successive reductions from the sociological to the biological, from the biological to the physico-chemical, etc. Several different points of departure have already been considered. One can in fact ask whether sociology and biology would not permit the direct induction of original logical forms and thence new mathematical structures. It is thought, also, that the mathematical forms elaborated in response to problems of physical chemistry have lost a little of their value and efficacy.

Two examples of tendencies in this direction can be cited; one is recent (this article was already partly written when I learned about it), the other dates farther back. The former example is from Piaget

(*Classes, Relations et Nombres*). He proposes the direct *union* between psychology and logistics which we have already referred to. "The deductive technique of logistics has acquired a rigorous precision, and the methods of the psychology of intelligence conform to the rules of experimental objectivity; why do these two sciences not collaborate as mathematics and physics have done?"

Thus the circle of the sciences would be closed.

The other example is from Vernadski, speaking to a learned society of Leningrad at a conference in 1930.

Vernadski begins by contrasting the Newtonian cosmos (with its infinite and homogeneous space where life plays no role) to the bio-human cosmos. His arguments at first are somewhat "empirical." If "the real content of science is the scientific work of living individuals," we must take into account the growing importance of the biological and anthropological sciences and the growing number of scientists who devote their efforts to them. Then again, serious "breaches" have been created in the Newtonian cosmos—changes bearing on the concepts of time, space, "causality," etc., produced by the new physics; the importance given irreversible processes in this new physics, etc.; but the crucial question is the following: may not biology be able to furnish concepts capable of transforming physics (instead of the reverse process)? We think we can cite an example of this in the researches on symmetry by Pasteur and Curie. In short, is physics a biology without life or is biology a physics plus life?

The mathematical-logical synthesis cannot be considered the adequate and necessary language of science; nor is it *one* of the sciences. In plain truth it is science itself. On the one hand, physical chemistry is tending toward an axiomatization on the order of geometry and algebra, which are hardly experimental any longer. In another direction we perceive the formation of a completely new mathematical logic qualified to absorb the domain of socio-biology (the so-called phenomena of consciousness and life.)

Thus, in whatever form one conceives the mathematization (or the logicization) of the different sciences, one cannot doubt the outcome of this development (a stage which scientific modesty feels obliged to deem provisional)—to wit, this very mathematization. Mathematics seeks itself through the different "sciences," as the sciences—science—seeks itself and creates itself through mathematics, which is at one and the same time the operational tool and the mode of perception.

37

THE ROLE OF MATHEMATICS IN THE DEVELOPMENT OF CONTEMPORARY THEORETICAL PHYSICS

by Louis de Broglie

ACADÉMIE FRANÇAISE. PERMANENT SECRETARY OF
THE ACADÉMIE DES SCIENCES

THE role played by mathematics in the development of physical theories is so well known that it may appear pointless to dwell on so hackneyed a subject. Progress in the various mathematical sciences has always gone hand in hand with advances in the great syntheses of natural philosophy. The concepts and methods perfected by geometers have always been of the greatest service to theoreticians dealing with physical phenomena, and have made their rapid advances possible. As just recompense for this, the study of physical phenomena constantly posed problems, the investigation of which led to the acquisition of new mathematical knowledge. To give only one example, let us take the most famous and important of them all, the discovery of infinitesimal analysis, for the honor of which Newton and Leibniz contended. As we know, this discovery played the decisive role in the rapid development of mechanics, astronomy and the first great theories of physics in the 18th century. And in this period, the study of astronomical and physical phenomena posed a great number of new problems whose solution considerably advanced the progress of mathematics itself. The theory of ordinary and partial differential equations, the calculus of variations and even a number of branches of algebra and geometry would not have advanced so rapidly if the Newtons, Eulers, d'Alemberts, Laplaces and Lagranges had not constantly faced the need to develop them in order to be able to make analytical interpretations of an ever-increasing number of facts, the observations of which and experiments upon which were almost daily adding to the treasure of human science. But all this is well known and has already been the subject of many studies and commentaries; it would no doubt be rather pointless to dwell on the mutual support that abstract mathematical research and the more concrete research of physics have given each other if this same substantiation had not recently won the attention of every philosophically inclined mind during the development of the latest theories of contemporary physics. These theories, the difficulty and boldness of which we know, could really not have arisen and developed without the groundwork of a formal mathematical structure and the analytical knowledge provided by the efforts of a long line of geometers who had thus, without knowing it, forged the instruments which were going to prove necessary for the progress of physical theories. And through the workings of the interaction which we pointed out above, the vast syntheses in our time of relativity and quantum physics in turn tax the ingenuity of the analysts with numerous difficult problems, most of which are still not fully solved. In these few pages we propose to show, with certain examples drawn from the development of contemporary science, how this collaboration and mutual interaction of mathematical and physical theories pursues its inevitable and productive course.

If one wishes to sum up the essential characteristics of contemporary physics in three words (perhaps an over-concise procedure), one could say: it is concerned with particles, relativity and quanta. This threefold description corresponds in a way to thesis, antithesis and synthesis. Indeed, contemporary physics has invoked the notion of discontinuity through the concept of particle which it has put at the base of a great number of interpretations. The simplicity and arithmetical character of the notion of discontinuity satisfy our minds, which are eager to decompose reality into its ultimate elements. But physics has not been able to rid itself of the opposing concept of continuity, which is the inevitable counterpart of discontinuity; it is due solely to this second concept, which conforms with another tendency of our minds, that we can utilize that admirable mental tool, infinitesimal analysis. Thus "field physics," an abstract expression for the continuous nature of the physical world, has never completely yielded to the physics of particles and has found its most complete and perfect expression in the theory of relativity. But arising unexpectedly, as if to bring about the difficult synthesis of continuity and discontinuity, of particle and field, there came upon the scene quantum theory and its extension, quantum and wave mechanics. The concepts of particle and field are preserved there, but with certain limitations and one might say in an attenuated form. At the same time a relationship of a statistical nature is established between them. This relationship, which creates the necessity for a fundamental revision of all the ideas previously admitted into theories of physics and entails the existence of "uncertainties"—which were spelled out in the celebrated inequalities of Heisenberg—does indeed constitute an astoundingly novel synthesis of the antithetical notions of continuity and discontinuity.

The particle theory of matter, field theory in its most perfect form, the relativistic, and quantum theory—these are the three great branches of contemporary theoretical physics. As happened in former times, these physical theories could not have made progress without constantly borrowing weapons from the mathematicians' arsenal, and they have posited many problems and suggested much research to mathematicians. The particle theory of matter has had recourse to results from the calculus of probability; field theory and relativity theory have called for the use of tensor calculus and the properties of Riemannian spaces; and finally quantum theory could not have been developed without the aid of the calculus of matrices and the theory of characteristic functions and characteristic values, the latter so closely related to the study of integral equations. This has resulted in a noticeable lag between the mathematics taught as preparation for the study of the physical sciences and the methods required for them today. Some currently advanced branches of mathematics which were not or still are not included in the mathematics curriculum intended for the physicist have become necessary for keeping up with advances in contemporary science; a reform of the curriculum has thus become indispensable, and it will take place sooner or later. Physicists of the generation of the author have had to learn entire chapters of analysis which had not been taught them in their youth in order to be able to stay abreast of their task, and this often burdensome effort has convinced them of the rapidity and the degree to which the "mathematical tools" employed by physical theories today have developed.

The successes of the particle theory of matter have led physicists to take up a mathematical discipline which before then had scarcely interested them: the calculus of probabilities. If matter consists of an immense number of particles in movement and capable of interacting with each other, its observable properties will be the statistical result of these movements and these interactions. If one supposes that the laws of classical mechanics apply to elementary material particles (which the founders of kinetic theories of matter did not doubt), it is evidently the calculus of probabilities that will permit one to go from the uncoordinated movements of these countless particles to the observable properties of the matter considered in the aggregate. The development in the 17th and 18th centuries of this very special mathematical discipline, the calculus of probabilities, was attended by many tribulations, encountering misunderstandings or sometimes even hostility on the part of many, often eminent, mathematicians, before being systematized through the genius of Laplace. It has found in particle physics a new and immense field of application. We know what prodigious successes have been won by the kinetic theories of matter thanks to the efforts of Maxwell, Clausius, Boltzmann and Gibbs. The interpretation of the properties of gases, permitting one to give a precise, dynamic meaning to the notions of pressure, temperature, specific heat, etc., the extension of these same ideas to liquids and solids, and over and above this, the remarkable interpretation in terms of probability, of that previously abstract and mysterious concept of thermodynamics, namely, the notion of entropy—all of these glorious conquests by theoretical physics at the end of the last century have rebuilt an entire section of physical science, allowing us to penetrate much more deeply into the ultimate meaning of the laws of thermodynamics. If certain fundamental results in the kinetic theory of matter can be obtained simply by resorting to rather elementary notions of the calculus of probabilities, this is not true when one wishes to investigate certain questions more thoroughly, or to attack the study of other more complex phenomena. The question of the persistence of velocities after molecular impact, as well as the problems raised by the investigation of fluctuations in density or energy, or by the investigation of Brownian movement or the non-stationary states of fluids, necessitate the employment of far more complicated and refined statistical considerations. Physicists who wanted to develop molecular and kinetic theories of matter were thus compelled to utilize all the resources of the calculus of probabilities, a thorough knowledge of which became indispensable to them. Conversely, mathematicians specializing in what had formerly been called game theory could not remain indifferent to the numerous and remarkable applications that their favorite science found in physics. Their cleverness was tested by numerous problems raised by these same developments in kinetic theories of matter. Thus the fundamental problem of Brownian movement or random motion gave rise to the mathematical theories associated with the Chapman-Kolmogoroff equation and the concept of "linked probabilities." This is a new example of the exchange of ideas and problems which enriches mathematics and physics at one and the same time. I shall dwell no further on this example, which concerns physical theories that are relatively old by now. I wish to devote my attention chiefly to more recent theories.[1]

The physical theories utilizing the notion of continuity and employing more or less explicitly the notion of field of magnitude were developed chiefly toward the end of the 18th and the beginning of the 19th centuries, in the era of the rise of the science of fluid motion or "hydrodynamics" and the science of internal solid motion or "the theory of elasticity." These naturally made constant use of the fundamental notions of infinitesimal analysis; here we see from the very beginning, in a more or less vague form, the mathematical operations and definitions which, when gathered together and systematized, were later to constitute vector analysis. A branch of this physics of continuity which investigated the disturbances propagated by waves in elastic media took on particular importance with Augustin Fresnel's splendid work in demonstrating the wave nature of light; these studies led him to represent light as similar to a transverse wave propagated in a particular elastic medium, the ether. Numerous results were achieved by the eminent mathematical physicists who contributed, toward the middle of the 19th century, to the development of the wave concept of light. All this work, by its very nature, belongs to the domain of "field physics," in the broad sense of the term.

Then field physics found a new area for expansion in the development of electromagnetism. In effect Maxwell's essential idea was to unify the known body of electromagnetic phenomena within a vast theory where fields and their step-by-step propagation would play the primary role. Thus, after Maxwell, electromagnetic theory became the most perfect example of a physical field theory. Even more than hydrodynamics or the theory of elasticity, it has accustomed physicists

[1] The article by Kahan (Part III, Book Three, pp. 104–117) extends the idea of this section. (Note by F. LL.)

to handling vector analysis and to making themselves familiar with mathematical operations which appear there quite naturally, such as gradient, divergence and curl.

Thus electromagnetic theory was led to consider problems whose solution required the introduction of advanced branches of mathematical analysis; examples of this are the theory of retarded potentials and the theory of the propagation of electrical disturbances along conductors. Lorentz's introduction of the notion of the electron in a form compatible with the continuity of field physics did not change the general aspect of electromagnetism in any way, while permitting it to interpret successfully a great number of important phenomena, such as the normal Zeeman effect and phenomena of electro-optics and magneto-optics.

However, we know that after Michelson's famous experiment and others analogous to it were carried out, electromagnetic theory in the Maxwell-Lorentz form began to encounter many apparently insurmountable difficulties. It required the emergence of the theory of relativity with its strangely novel concepts to enable the physicist to escape from this impasse. There is no need for us to summarize the vicissitudes encountered by physical theory in this very curious period, nor to sum up the solutions proposed by the theory of relativity. What we wish to stress are the mathematical tools which Einstein and his followers were led to use in order to carry out their lines of reasoning, and which were practically unknown to physicists until then. In a sense the theory of relativity can be regarded as the crown of classical field physics, because it seeks to represent in one stroke the entire body of physical facts, past, present and future, within the framework of a four-dimensional continuum, space-time, which achieves a fusion of sorts between space and time. Although the unidirectional flow of time and the entire experience of our psychological life, as well as certain features of quantum theory, advise us that a complete integration of space and time undoubtedly does not conform to basic physical reality, it is certain that the representation of phenomena in space-time essayed by the relativists has revealed kinships between physical magnitudes that were formerly hidden and has permitted us to represent certain of their relationships in an elegant and precise form that has often been a valuable guide to researchers in achieving new discoveries. The mathematical tool that has proved necessary and essential for bringing the concepts of relativity theory into sharp focus has been tensor analysis, which is a generalization of vector analysis;

from an abstract point of view, both are no more than a branch of the algebraic theory of linear substitutions. To be sure, many branches of physics, notably the theory of the elasticity and optics of anisotropic media, had made use of the notion of tensors before the development of relativity, but they had not done so in a systematic fashion extending to the full the general method employing these mathematical entities. Before the theory of relativity, physicists had quite often carried out tensor operations without even knowing it, just as Monsieur Jourdain had been speaking prose. Afterward physicists became aware that it was essential for them to know and be able to handle tensor analysis, and mathematics came to their aid by giving this analysis a rigorous form and by fixing its laws. Later it became clear that all the branches of physics (and not only the theory of relativity) could make very profitable use of this analysis, and many of the chapters of classical physics have been as it were translated into this new language, whose elegant and condensed form illuminates relationships and analogies which the previous forms of presentation masked or at the very least could not display so clearly.

Tensor analysis, supplemented by the classical methods of mathematical analysis with which physicists had long been familiar, had sufficed for developing the original theory of relativity, known as "special relativity," where only rectilinear and uniform relative motions of the systems of reference are considered, but not accelerated motions. However, a theory so restricted could not suffice because it was patently incomplete. Its concepts had to be generalized so as to take into account accelerated motions and the absolute nature of the accelerations. This is what Einstein achieved in 1916 in his general theory of relativity, which introduced, as we know, a most novel and curious interpretation of gravitation. The essential idea of this ingenious theory was the declaration that space-time is a Euclidean continuum analogous to a hyperplane only in those of its regions situated far away from any material body endowed with mass. The proximity of a material body has the effect of giving the regions of space-time a curvature, so that locally space-time is the four-dimensional analogue of a curved surface and no longer of a plane. The phenomena of gravitation result from the existence of this curvature, and gravitation thus receives, if not a real explanation in the ôlder meaning of the term, at least an interpretation.[2] Without entering

[2] On this subject, see the article by Thiry, Part I, Book Two, pp. 143–151. (Note by F. LL.)

into the details of this beautiful but difficult general theory of relativity, it is easy to understand that its development requires it to lean
upon the analytic investigation of curved spaces of any number of
dimensions, which generalizes the investigation of curved surfaces in
our three-dimensional space. The aforesaid investigation was made
well before relativity theory by 19th-century mathematicians. Gauss,
in particular, introduced the systematic use of curvilinear coordinates
and thus laid the foundations of the general theory of surfaces.
Riemann generalized the use of curvilinear coordinates and undertook
a powerful analysis of metrical concepts in a continuous space of any
number of dimensions, creating with this work the justly named
theory of Riemannian spaces. These beautiful mathematical syntheses, of an advanced level, were well known to mathematicians but
far less so to physicists—even to theoretical physicists, who only in
exceptional cases used certain fragments. The general theory of
relativity obliged physicists to know these syntheses well and to apply
them frequently. Advocates of general relativity regularly made good
use of the mathematical studies on Riemannian spaces, particularly
those done by the Italian school; Levi-Civita's concept of parallel
displacement rendered especially great service in this area. And as
always, by showing interest in the theory of polydimensional spaces
from the point of view of applications, physical theory inspired
mathematicians to continue and deepen the study. It will suffice to
cite as examples the remarkable work of Elie Cartan and Hermann
Weyl, who considerably enriched this branch of the science and were
constantly inspired and guided by developments in relativity theory.
The desire to fit the electromagnetic field into a geometric scheme
analogous to the one that had succeeded in interpreting gravitation,
but of a more general nature, led to the various endeavors known as
unified theories of gravitation and electromagnetism; these theories
were due principally to Weyl, Eddington, Einstein and Kaluza. Some
of them replace the space-time of four dimensions by a multiplicity of
five dimensions, others turn it into a continuum having more general
properties than those of Riemannian spaces. Although these unifying
theories have not up to now succeeded in representing the true nature
of the electromagnetic field very satisfactorily and although we must
certainly wait to hear the quanta speak their piece in this affair, the
theories have nevertheless inspired an entire movement of ideas that
has been very profitable for both theoretical physics and pure mathematics.

This rapid examination of the various, successive stages in the theory of relativity enables us to appreciate the considerable role which the most advanced branches of mathematics have played in the most memorable advances of contemporary physics. This idea will be further accentuated when we take a glance at the development of the most astonishing conquest the physicists of our time have achieved: quantum theory and its extensions.

Those who have followed the evolution of quantum theory know that at first it consisted of the introduction of certain very surprising arithmetic discontinuities into the study of small-scale particle motion and radiation. This first formulation, known as the old quantum theory, was created by Planck in a flash of genius in order to find the exact black-body laws; Einstein transposed it to the domain of radiation with his astonishing concept of light quanta; and finally, Bohr used it in his model of the atom, universally known to all scientists today. In this original form quantum theory did not resort to using advanced mathematical ideas with which physicists were little familiar. To be sure, the theory attached great importance to the mechanical notion of action, which had been introduced into science earlier and in a rather peculiar fashion by the celebrated Maupertuis; this rather abstract and not very intuitive notion is, with respect to these qualities, somewhat analogous to entropy, with which it may even have deeper, but still imperfectly known, affinities. But to sum up, the idea of action was well known to all who had seriously studied analytical mechanics. The arithmetic relationships which were introduced into the old quantum theory were also confined to very elementary ones. However, this aspect of the situation began to change in 1916 when Bohr introduced his principle of correspondence into quantum theory. A precise statement of this principle necessitated the introduction of theories of analytical mechanics, such as the theory of angular variables, which though classical for mathematicians were in general quite unfamiliar to physicists. The development of this last phase of old quantum theory would have been impossible, or at least much slower, without the long and laborious labors of the authors of celestial mechanics, a subject to which the greatest mathematicians from Laplace up to Henri Poincaré have not disdained to add their contribution. The calculations of Sommerfeld and other physicists who extended the investigation of Bohr's model of the atom by applying a broader idea of the quantification of motion also made constant use of the most advanced portions of analytical mechanics, notably Jacobi's well-known theorem. Mathematical instruments forged long beforehand by eminent mathematicians were already rendering inestimable service to quantum physicists.

But this was still nothing. The appearance of the new quantum theory in its twofold form of quantum mechanics and wave mechanics was to involve physicists in a still greater utilization of mathematical concepts and methods in which they had shown until then very little interest. The quantum mechanics due to Heisenberg makes constant use of the mathematical entities called "algebraic matrices." Their properties were long known to the mathematicians who encountered them in the study of linear transformations. They entered more or less imperceptibly into a number of physical theories where linear transformations play an essential role and where matrices and tensors often appear side by side. But physicists had never (we believe) made explicit use of matrices, with whose properties they had little familiarity. However, first quantum mechanics and then wave mechanics could not avoid making constant use of them. Physicists were therefore obliged to study the theory of matrices, and they were sometimes even compelled to extend this theory by introducing new notions such as the derivative of a matrix with respect to another matrix. Then, having mastered the techniques of matrix analysis, physicists became aware of the fact that it could render them great service even outside the domain of quantum theory. Problems of great importance today in electrical engineering, but without any quantum features, for example those relating to quadripoles and electric filters, can be handled rapidly and elegantly thanks to the systematic use of matrices. This has been demonstrated especially by Léon Brillouin; electrical engineers have profited greatly from his great skill in handling the methods of quantum theories.

Wave mechanics was created some months before quantum mechanics and was suggested by considerations which were very different and on the surface more intuitive. But we know today that, judged from a purely formal point of view, the two theories, despite their very dissimilar appearance, are not essentially distinct; they strike one as being two translations of the same fundamental idea into two different mathematical languages. While quantum mechanics employs the formalism of matrices, systematically and a priori, to represent the physical magnitudes characterizing the motion of elementary particles, wave mechanics represents the state of these particles by certain wave functions analogous to those which had been used in the study of the vibration of material bodies and those which were employed for a century in the wave theories of light. It was only secondarily that wave mechanics with the aid of these wave functions defined certain magnitudes having the properties of algebraic matrices; but when Schrödinger showed that these matrices are identical with those that Heisenberg had introduced directly into his quantum mechanics, the fundamental identity of the two new mechanics turned out to be established.

The essential problem of all the quantum theories is the determination of the quantified steady states of material systems at the atomic level; this problem appears in wave mechanics in a form that is particularly interesting from a mathematical point of view. As a matter of fact, in this mechanics the motion of elementary particles is associated with a wave propagation, and in every material system this wave propagation is governed by an equation of propagation which can be considered a generalization of the equation of vibrating strings, but whose coefficients depend on the nature of the particles constituting the system and on the fields of force to which they are subject. This quite naturally leads wave mechanics to define the quantified steady states as corresponding to standing waves (in the classical sense of the term) whose source can be the system. Mathematically their determination amounts to finding the steady-state solutions of the propagation equation satisfying certain boundary conditions imposed by the very nature of the question. Now this is a problem of a type well known to physicists and mathematicians since d'Alembert's and Euler's work on vibrating strings. It had been encountered long before in the theory of elasticity with the study of vibrating strings, rods and diaphragms; in acoustics with the problem of sound chambers and tubes; in electricity with the determination of the electromagnetic oscillations that radio antennas and enclosures bounded by metal walls are susceptible to, etc. In the simplest cases, for example that of a vibrating string fixed at both ends, or the case of a plane membrane rectangular or circular in shape and with fixed perimeter, the mathematical solution of the problem is easy to obtain; but here already a question comes up that is more difficult to prove: that any vibrating state whatever of the system under consideration can be represented by a superposition of simple steady states conveniently chosen. It is very difficult to give a proof of this very general statement; the problem did not cease to engross mathematicians from the time of Euler until the discovery of integral equations. It ties in with the use of decompositions into series and into Fourier integrals which correspond exactly to the representation of any given vibrating state by a super-position of steady states in the simple case where these steady states are given by trigonometric functions. Encountered by Euler in the middle of the 18th century, this type of development was given particular attention by the illustrious mathematical physicist Joseph Fourier in his *Analytical Theory of Heat*, an outstanding work which remains one of the most beautiful monuments created by French science in the magnificent period of its flowering at the beginning of the 19th century. Investigation of the phenomena of heat conduction led Fourier to determinations of steady states completely analogous to those encountered in the theory of vibrations, and these led him to introduce systematically the development in series or integrals of trigonometric functions, which justly still bears his name.[3]

Considered in all their generality, the problems of steady states pertaining to vibratory and heat phenomena, among others, have gradually given rise to a general mathematical theory which today is usually called the theory of characteristic functions and characteristic values of differential equations (or partial differential equations). It is concerned with proving the existence of steady states and determining the form of the functions which represent them. It is called upon to examine the following question in its most general form, namely, whether it is always possible to represent any given solution of a differential equation by a sum of certain of its characteristic functions, this representation corresponding to the representation of any given vibrating state by a superposition of steady-state vibrations. This question leads to a study of the types of developments in series or integral expansions which generalize those for which the Fourier series and integrals were the prototypes. For a long time there were only partial solutions to all these difficult problems, and the existence of characteristic functions and characteristic values and the correctness of corresponding series expansions had to be proved for each particular case. This was still the situation at the end of the past century when Henri Poincaré, in a book of the utmost interest, submitted to critical study the problems relating to the conduction of heat which had become classic since Fourier. Shortly after this, however, the development of the beautiful theory of integral equations, due

[3] This entire paragraph and the following one are to be compared with the article by Janet, Part III, Book Three, pp. 94–103. (Note by F. LL.)

chiefly to the work of Fredholm, was to cast a totally new light on everything connected with this problem, and was to furnish mathematicians with powerful methods for tackling general proofs which it was previously not possible to carry out. Investigations concerning function spaces, whose principal author was the great German mathematician Hilbert, furnished an appendage to the theory of integral equations, completing and illuminating it.

In particular, these investigations have revealed the parallels existing between the theory of characteristic functions and characteristic values and that of algebraic matrices, parallels that are the fundamental reason for the identity of wave and quantum mechanics. Thus, an entire corpus of mathematical work, originating in certain problems posed by classical physics but not followed up very seriously by physicists themselves, has today become indispensable knowledge for all who would investigate the new quantum theories; in turn, the investigation of these new theories has already furnished and will continue to furnish mathematicians with many new problems to solve. The mathematical problems presented by the quantum theories are far from the simplest from the point of view of analysis; they often correspond to integral equations having singular kernels, infinite domains and continuous spectra. They are just those exceptional cases which Fredholm's general theorems do not apply to. The rigorous proof of the existence of characteristic values or of the correctness of series expansions of characteristic functions therefore requires special examination. Thus the new quantum theory provides mathematics with many interesting problems to study.

The rise of quantum and wave mechanics required more than the introduction of matrix analysis and the theory of characteristic values.

Mathematicians had long appreciated the great importance and vast area of application of another large branch of mathematics, group theory. This now became the useful and sometimes indispensable tool of the quantum physicist. For many years physicists had almost totally ignored group theory. To be sure, the penetrating work of Pierre Curie on the symmetry of physical phenomena had implicitly introduced certain conceptions belonging to this theory, but the utilization thus made was very incomplete and partial. In wave mechanics, as the work of Hermann Weyl, Van der Waerden and Slater in particular has shown, the results of group theory applied to the investigation of atoms having a large number of electrons permits one to quickly obtain proofs which would be very difficult or often even impossible to obtain otherwise. The detailed calculation of wave functions for a system containing several components quickly becomes impossible when the number of components increases. But considerations of symmetry that can be applied in a systematic form in group theory, permit the discovery of certain global characteristics of wave functions from which one may deduce many properties of the system under investigation, without having to go through the exact determination of these wave functions. In particular, some very important general statements were thus obtained relating to the global kinetic moments of atoms or molecules with large numbers of electrons and to the rules of selection which permit one to predict which quantum transitions these atoms or molecules are capable of executing. Thus a very advanced and very abstract branch of mathematics has found a wonderful field of application in the rapid prediction of phenomena which are of the greatest importance to physicists.

We have seen mathematics providing necessary and effective assistance in the advance of the great theories of modern physics. Today the same fact may be observed to an equal degree in the development of many more specialized branches of physics. I shall give only some examples relating to electrical theory.

As we know, the concept of a complex or imaginary number arose in the investigation of algebraic equations, where it permitted every equation of the nth degree to be thought of as always having n real or imaginary roots. Developed at the end of the 18th century, this concept was to furnish Cauchy with the necessary element for creating his beautiful theory of analytical functions, the basis of all modern analysis. But for a long time this notion of imaginary quantity, which seemed abstract and somewhat mysterious, was barely used by physicists; little by little, however, their disinclination toward it waned, and we see them making use of imaginary quantities, notably in the study of vibrating phenomena and in the wave theory of light, where the substitution of imaginary exponentials for the corresponding trigonometric functions permits one to carry out calculations with much more speed and elegance. Quite naturally, this use of imaginary quantities was extended to the investigation of alternating currents, where it was not slow to render great service; thus arose today's widely used methods of calculation, known to electrical engineers as symbolic methods. Developments of these methods have not ceased, as can be seen by referring to the recent history of modern electric theory, and in particular to the fine work of the late Jean Fallou; they provide elegant condensed expressions, especially for effective power, with watts or without, that come into play in alternating-current circuits.

The circulation of electric currents in circuits composed of resistances, inductances and capacitances rests on simple laws. However, as soon as one considers the circuits of a somewhat complex structure, the investigation quickly becomes complicated, especially if one does not want to confine himself to normal loads, but also wants to predict transient loads. The problems thus posed are very important in the practical applications of electricity and have been the subject of many studies in the past fifty years. The most original were certainly those of Oliver Heaviside, who created a totally new method of calculation for investigating electric circuits in general, a method known today as "Heaviside's symbolic calculus." Proceeding more by intuition than by rigorous reasoning, Heaviside guessed rather than proved most of the results he obtained. It required the efforts of many scientists to place Heaviside's method on a truly rational foundation. Carson seems to have made the biggest contribution in this area by linking all the English physicist's results to the properties of a certain integral called "Carson's integral."

For some years a considerable number of works have been devoted to Heaviside's method; while electrical and radio engineers were being more attracted to it every day and were varying its applications more and more, the mathematicians were consolidating its foundations and in the course of tying it in to well-known analytic concepts such as the Laplace transformation or Fourier integrals, employed it to handle various problems of analysis and even problems in the theory of numbers. Rarely has it been so evident how a new method can simultaneously be of interest in the most advanced research in pure science and the most important problems in applied science.

The technical utilization of short-length electromagnetic waves (of the order of a decimeter) is in full development today. This leads to the study of the propagation of such waves in diverse circumstances, particularly within the interior of metal tubes called "wave guides," the shape and dimensions of which may vary. These problems of wave propagation call once more upon the theory of characteristic functions and characteristic values, whose great importance in all branches of physics we reviewed above. Their complete solution makes use of various functions discovered and studied by mathematicians of the past two centuries, such as Bessel, Laplace and Mathieu functions, and the polynomials of Legendre, Sonine and others. Thus a branch of the science of radio communications, one of the richest in possible future applications, is constantly using methods made available to it through the long and patient work of the pioneers in mathematical analysis.

If we also recall that the theory of matrices created by algebraists is now serving electrical engineers, after having passed through the hands of the quantum physicists, we shall be able to appreciate how much a science that is directed principally toward practical objectives, as is electrical engineering by its very nature, is beholden for part of its recent progress to the most advanced mathematical research. All the physical theories, from the great general theories which are developed in an atmosphere of disinterested research with no regard for utility, up to the more special theories of the various branches of applied physics and those more concerned with immediate results, make greater and greater use of the results and methods made available to them through the work of successive generations of mathematicians. This reveals the unity in the work accomplished by various kinds of scientists who often appear committed to very different points of view; the final convergence of their efforts is thus vigorously affirmed.

38

HARMONICS AND SPECTRA, VOLTERRA'S IDEAS, FREDHOLM'S EQUATION, HILBERT SPACE, CLASSICAL PHYSICS AND MODERN PHYSICS

by Maurice Janet

PROFESSOR AT THE SORBONNE

THE sounds which strike our ears come from the infinitely varied types of vibrations of the bodies surrounding us. These various types may coexist to a greater or lesser degree, but it may happen that one of them—a simple one—clearly dominates the others. In any event, in order to get a clear look into phenomena of such complexity it will be a good idea to examine first of all the case of a material system possessing a single degree of freedom, i.e., one whose configuration at each instant can be defined by a single number, such as the position of a point on a given straight line; as is natural in a beginning study of elastic phenomena, we accept as true that potential energy is proportional to the square of the distance from the position of equilibrium (displacement), and moreover that the sum of potential and kinetic energies is conserved; we thus perceive the law of motion: sinusoidal oscillation, where the square of the frequency is directly proportional to the "rigidity" and inversely proportional to the "inertia"[1] (the two coefficients of rigidity and inertia being the constant quotients respectively of the potential energy divided by the displacement

---

[1] In other words, the point considered is simply subject to an attraction toward a fixed center that is proportional to the distance. Its "coefficient of inertia" is its "mass."

squared over 2, and the kinetic energy divided by the velocity squared over 2). The same simple law naturally holds for many phenomena other than those involving material elasticity: If a condenser is discharged through a coil of heavy conducting wire of negligible resistance, the electric charge of each plate is a sinusoidal function of time; for the factors of rigidity and inertia we simply substitute the inverse of the capacity of the condenser, and the coefficient of self-inductance of the coil.

Let us now imagine a material system with several degrees of freedom, three for example. To specify its configuration, it will be necessary and sufficient to give three numbers; an example would be a free material point in space. Let us accept that its potential energy is a positive homogeneous function of the second degree in the coordinates, and that here too "the system is conservative." Elementary principles of mathematical analysis permits us to see that its most general motion may be thought of as the *superposition* of three simple oscillating motions with fully determined frequencies. In one of these simple motions the three coordinates vary in synchronization with the frequency (only the mutual relationships of the three amplitudes depend on how the system is constituted; the rest, the coefficients of amplitude and phase, depend upon the initial conditions). Determining the three "*fundamental frequencies*" is the key to the problem here. Important "extremal" properties can serve to characterize them. Let us say merely that finding the three fundamental frequencies of such a vibrating system can be compared with finding the axes of an ellipsoid.[2] Entirely analogous results can be immediately stated for a material system possessing any number n of degrees of freedom.

When a finite number of parameters no longer suffices to characterize the material system under study, a new and decisive step must be taken. To compensate for this, we shall continue to accept the *linearity* of the relationship between "force" and "displacement" that is implied in the earlier hypotheses. This *linearity* will play an important role in what follows.

To define the position of a vibrating string in a plane at a given instant, a function of a single variable is all that is necessary. When we are dealing with strings, membranes, organ pipes or any other

---

[2] Algebraically, in both cases we are dealing with the simultaneous reduction of two given quadratic forms to the "normal" forms, i.e., forms devoid of "mixed terms"; the two forms correspond here to the potential energy and the kinetic energy of the system.

vibrating system, then our unknowns will be functions of one or several spatial variables. And we shall pass from a finite number to an *infinite set* of basic frequencies. This is what the celebrated example of the homogeneous vibrating string fixed at both ends illustrates (Daniel Bernoulli, 1741). The basic frequencies are successive multiples of each other; the corresponding sounds are the "harmonics" of the fundamental tone. The superposition of simple solutions gives the general solution of the problem; and this fact can be considered the starting point for all the burgeoning discoveries we are going to discuss (Fourier, 1811). Suppose again that we are dealing simply with a perfect vibrating string but we no longer assume its homogeneity; it is going to be necessary to show its synchronous modes of vibration— what physicists call *partials*—and above all, the frequencies of these partials. The latter have, understandably, often been confused at first with the harmonics we found previously.[3] The laws of classical mechanics lead to the statement that a certain linear differential equation (without a "second member") which depends upon a parameter λ, admits of one solution, not identically zero, satisfying certain boundary conditions. Such a problem is possible only if λ has one of the values of a certain infinite discrete sequence, depending simultaneously on the equation and on the conditions imposed; knowing these remarkable values of λ leads immediately to the basic frequencies. Whether it was dealing with mechanical or electromagnetic vibrations or with any number of other problems, physics during the 19th century found itself constantly being driven back to questions of this kind.

The mathematical proof of the existence of the infinite series of "fundamental values" ("characteristic values" or "eigenvalues") cost great efforts; we need only enumerate some names and dates to realize this. After the first efforts of Sturm and Liouville (1838), Schwarz succeeded in proving the existence of the first in 1885, Picard the second in 1893 and Poincaré all of them in 1894. It was reserved for Fredholm and Hilbert to present a so to speak perfect formulation of the entire theory by mastering it from the point of view of "integral

---

[3] In Diderot's *Le Neveu de Rameau* (1762) appear these rather mysterious lines: "Provided the bells of his parish continue to ring the twelfth and the seventeenth, all will be well." What is this riddle? Why these numbers and not others? In the ordinary scale the notes corresponding to the double, triple, quadruple and quintuple of the frequencies of the tonic (or note of rank 1) have ranks of 8, 12, 15 and 17. Rejecting the *eighth* and *fifteenth* as too banal (the octave and double octave), the character in question naturally thinks of the *twelfth* and the *seventeenth*.

equations." In 1896 Volterra had given the general method of solving such equations for a simple case; for several years previously he had been systematically employing in his research the fruitful method of going from the finite to the infinite, from the discontinuous to the continuous.[4]

There is a preliminary problem which we can put in the following form: Given the composition of a material system and the boundary conditions, how does knowing the elastic forces, after all, allow one to proceed to knowledge of the displacements? Above all, if the forces differ perceptibly from zero only in a small region A, how does one deduce the displacements in another region B? It is the "coefficient of influence" or "Green's function" which permits us to answer the question; for an arbitrary distribution of forces, we shall need only to multiply the "force" at A_i by the coefficient of influence corresponding to A_i, then add all the results obtained in order to have the "displacement" at B. The result is that the preliminary problem is solved by an "integral" extended over a determined field, involving a product of two functions, one of which, the coefficient of influence, contains the point B. The coefficient of influence has a remarkable property of symmetry: It does not change when A and B are permuted.[5]

Supposing the preliminary problem, investigating "Green's function," to be solved, let us return to the differential equation containing the parameter λ which we discussed above. Briefly, this equation would give, as a function of λ and the "displacement," the expression for the elastic force corresponding to the case of these special motions that we are seeking to determine. By using this same expression in the integral we have just defined, we will obtain an equation where the unknown function "displacement" will figure linearly *through its value at B first of all*, but also under the integral sign through *its values at all the points* A_i of the domain under consideration. It is the linear integral equation "of the second kind," with fixed limits, with which Fredholm's

[4] Integral equations are already to be found in Laplace (1782) and Fourier. Abel (1823) dealt with a celebrated integral equation. Liouville (1837) and Neumann (1870) gave a solution by means of successive approximations. A propos of Volterra's ideas, we should recall Cauchy's method (published in 1844 and since perfected by Lipschitz) for establishing the existence of solutions of ordinary differential equations, a method of which Euler had already provided a glimpse.

[5] There is a well-known illustration of this fact: In the case of a girder kept horizontal in any desired manner, the bending produced at a point B when a weight is attached at A is the same as that which would be produced at A if the same weight were attached at B.

name is henceforth associated. Furthermore, it is an equation "without a second member" and with a "symmetric kernel."[6] This symmetry has an obvious algebraic source; the study of a quadratic form in n variables leads naturally to associating it with a *symmetric* square table of n rows and n columns. Essentially, Hilbert[7] intended to generalize the theory of quadratic forms to the case of an infinite number of variables; he reaped an amazingly abundant harvest of results. At the same time, Fredholm[8] was providing definitive theorems for the integral equation of the second kind, with fixed limits, with or without symmetric kernel and with or without second member. He makes some general hypotheses about regularity as concerns the "kernel," but it does not much matter now whether one or several variables were involved, whether we were dealing with an ordinary differential equation or a partial differential equation, or whether or not the problem had a physical origin;[9] the method of solution is uniform, and the results are a direct generalization of the elementary theory of systems of linear equations. We distinguish two different cases: 1. λ does not reduce to zero a certain entire function $D(\lambda)$ whose expression is given by Fredholm, and the equation has one and only one solution (the "meromorphic" function of λ); 2. λ does reduce D to zero, and the equation is possible only if the "second member" satisfies certain readily formulated linear conditions, finite in number. It may happen that $D(\lambda)$ has no zero. It is a remarkable characteristic of symmetric kernels that they always give rise to at least one singular value; moreover the singular values are all real, and finite in number only for kernels of a particular, simple form: the sum of a finite number of products of functions of one or the other variable (Goursat's kernels).[10] The infinite set of singular values of λ naturally

[6] The "kernel" of a linear integral equation being this function as given by two points which characterize the equation in essence, just as a certain two-dimensional table of coefficients characterizes an ordinary system of linear algebraic equations. The kernel is called "symmetric" if it depends symmetrically on the two points considered.

[7] Hilbert: *Grundzüge einer allgemeinen Theorie der linearen Integralgleichungen* (1912; collection of papers published from 1904 to 1910.)

[8] Fredholm: *Acta Mathematica* (1903), published in a collection; a first work on the subject had been published in 1900.

[9] The celebrated boundary-value problem associated with the Laplace equation, known as Dirichlet's problem, reduces to a Fredholm equation if one simply looks for a "double layer" distributed on the given boundary.

[10] This is a very important case precisely because it allows us to discover all the essential results of the theory, thanks to a passage to the limit (using the

furnishes the set of fundamental values encountered in the physical problems mentioned above; these are the "characteristic values" or "eigen-values" or yet again the "spectrum" of the given kernel. For each of these the equation without a second member has a finite number of linearly independent solutions; the infinite set of solutions thus obtained can be considered an "orthogonal and normal" set. Any function satisfying only certain general conditions of regularity can be developed in series of these functions. The Hilbert-Schmidt theorem includes as a special case the development in a trigonometric series and the more or less isolated developments found after Fourier's famous discovery. Moreover, the fundamental values and corresponding solutions allow us to find the kernel; under certain hypotheses we obtain its canonical form or "spectral decomposition"; each term of the infinite series obtained is the product of the values taken on by a fundamental function for one or the other variable, divided by the corresponding fundamental value.

But the most profound of Hilbert's results concern some cases where Fredholm's theorems would be powerless. Hilbert's investigation starts with the study of a space each of whose points is defined by an *infinite set*[11] of coordinates such that *the sum of their squares converges*. His whole theory rests on his choice of this condition.[12] Quadratic forms with an infinity of variables, the sum of the squares of whose co-efficients is convergent (the forms known as "completely continuous"), can be reduced to a canonical form entirely analogous to the one we encounter in dealing with a finite number of variables. The sum of a finite number of terms is simply replaced by an infinite series, and the application to integral equations gives the "spectral decomposition" mentioned above. Completely continuous forms are "bounded" in the sense that the finite forms made up of their first terms ("truncated" forms) remain less in absolute value than a fixed number which is independent of the number of terms retained, when the sum of the squares of the coordinates of the variable point remains less than unity, but the reciprocal is not true, and the investigation of *bounded* forms that are not completely continuous leads Hilbert to a canonical

notion of an equi-continuous *family* of functions). (Cf. *Mémorial des Sciences Mathématiques* (1941), fasc. 101 and 102.)

[11] Going from the finite to the infinite, a procedure employed by Fredholm and Volterra, was at the same time a passage from the discontinuous to the continuous. Hilbert goes from a *finite series* to an *infinite series*.

[12] See the article by Dieudonné, Vol. I, p. 307. (Note by F. LL.)

form of a totally new kind; to the *series* previously found, an integral[13] must be added: to the spectrum of points, a continuous spectrum must be added (the spectrum of points could then, incidentally, disappear completely).[14] These beautiful discoveries shed a strong light on the theory of "singular" integral equations; and the previously known results for Fourier's integral in turn illuminate this new theory.

Let us return to (Fredholm's) regular integral equations and again let us limit ourselves to symmetric kernels. The "fundamental values" are characterized by important extremal properties which have already appeared in the study of very simple vibrating systems possessing several degrees of freedom. If we allow only one degree of freedom to such a system by adding linear relationships, the square of its frequency appears as the ratio of two quadratic forms whose coefficients in the denominator and numerator come respectively from the general expression for the kinetic and potential energies of the given system. The "fundamental" motions are those which make such an expression "stationary." In his beautiful *Theory of Sound*, Rayleigh had considered the analogous relationship for the case of a continuous system such as a vibrating string. Hilbert and his school, in particular Courant, have shown how this fact and its analogues permit us to prove, in the clearest and most direct way, not only the existence of the singular values of symmetric kernels, but also their most important properties.[15]

These properties clearly reveal some general truths about numerous physical applications: adding a constraint, reducing the factors of inertia, increasing the factors of rigidity, can at the most augment but never diminish the heights of the successive partials of the vibrating system.[16] These same properties permit the derivation of the law of the *asymptotic distribution* of the characteristic values. Let us suppose a homogeneous distribution of the factors of "rigidity" and "inertia"; the expression for the number of characteristic frequencies less than a

[13] In Stieltjes's sense.

[14] This situation appears, for example, in the form $x_1 x_2 + x_2 x_3 + x_3 x_4 + \cdots$, the spectrum of which consists of the total number of real values outside the interval $(-1, +1)$.

[15] Courant and Hilbert: *Methoden der Mathematischen Physik*. (First ed. 1924, 2nd ed. 1931.)

[16] Thus, if one point of a vibrating string is fixed, the frequencies due to the "normal" vibrations of one or the other part constitute in all a set, the values of which *come between* those of the original set. (Lord Rayleigh: *The Theory of Sound*, vol. I, 2nd ed. (1929), p. 22, and Van den Dungen: *Les problemes généraux de la technique des vibrations*, 1928, p. 83.)

given quantity does not depend on the form of the domain being considered, when the said quantity becomes large; it depends only on the volume of the domain—a simple result which Lorentz had foreseen in connection with radiation theory.[17] Finally these are the properties that serve as a foundation for the methods of the numerical techniques developed to study characteristic values (Ritz),[18] and also for the methods used by engineers to solve problems concerning the effective use of materials.[19]

But it is remarkable that in recent years the most dazzling applications have been in a totally new direction.

In the era when Volterra, Fredholm and Hilbert were active, physics proceeded from discovery to discovery. The laws of blackbody radiation led Planck to the hypothesis of quanta of energy; shortly thereafter, this hypothesis, along with Einstein's theory of "photons," resulted in a kind of return to the corpuscular theory of light once advanced by Newton, which, in consequence of Fresnel's and Maxwell's memorable discoveries, had been abandoned in favor of the wave theory. For a brief moment Poincaré could wonder whether physics was going to have to give up using differential equations. For about twenty years it was necessary to use one or the other of these seemingly contradictory theories, depending on the experimental phenomena one wanted to explain. Light seemed to have a dual and completely paradoxical nature. Almost simultaneously two different and superb conceptions for resolving these difficulties appeared on the scene: One, brilliantly promulgated by Louis de Broglie (1924) and considerably extended by Schrödinger (1926) finds the solution by giving matter itself a dual nature, both wave and corpuscular. Into the simple investigation of the motion of a material point, this first method introduces a partial differential equation of the type to which the study of strings, membranes and other complex vibrating systems of the "old mechanics" had already led. And as one might expect, the discontinuous set[20] of the corresponding "characteristic values"

[17] Lorentz: *Physikalische Zeitschrift*, 1910, p. 1248.

[18] Ch. Kryloff: *Les méthodes de solution approchée des problèmes de la Physique mathématique.*

[19] It is not only with reference to vibrating phenomena that we get questions relevant to these same theories; one example is the problem known as the "buckling" of a longitudinally compressed rod, whether motionless or in rotation around its axis.

[20] In important cases a continuous spectrum (Schrödinger) must be added to the discontinuous spectrum, a remarkable illustration of Hilbert's general results.

yields the set of privileged *levels of energy* which experiment had revealed in the atom.[21] The other conception, due to Heisenberg (1925), is at once more abstract and yet closer to experience. It seeks to introduce into its calculations only those magnitudes which correspond to a possible physical measurement; it therefore finds it necessary to study those infinite square tables of numbers called "matrices," and on the whole to study the linear transformations in Hilbert space. And here again the problem involves "characteristic values," in this case the characteristic values which determine energy levels. At first these two theories seem to be very different. In reality they are equivalent (Schrödinger, 1926). And, with von Neumann,[22] one can see the essential reason for this equivalence in a remarkable fact[23] discovered by Riesz in 1907: the abstract identity of Hilbert space and that of "summable square" functions. It is interesting to note that before the discovery of such a result could be made, it was necessary for the notion of the integral to be clarified; Lebesgue did this in 1902.[24]

Fredholm's equation, and with it linear integral equations of various kinds, Hilbert space, and other more or less analogous spaces, have been the subject of a considerable number of studies since the beginning of this century, with the Lebesgue integral a current tool. Carleman[25] has made great strides in the study of singular integral equations; von Neumann, starting with an axiomatic definition of Hilbert space, has obtained an intuitive method leading readily to numerous results.[26]

The problems which first led to integral equations arose in the study of nature; the study of "acoustics" and "vibrations" was a precious guide.

But mathematicians quite properly made wide use of their freedom in developing and enlarging the theory. Analytic methods and synthetic methods, algebraic analogies and geometric analogies were employed by turn. During this period increasingly numerous and

[21] The simple problem of the "linear harmonic oscillator," with which we commenced this study, led to the consideration of the infinite set of orthogonal functions to which Hermite's name is attached.

[22] Von Neumann: *Mathematische Grundlagen der Quantenmechanik*, 1932.

[23] Fischer pointed this out simultaneously in a somewhat different form.

[24] On this subject see Perrin's article on Lebesgue (Part II, Book Two, pp. 298–303), as well as the relevant passages in Valiron's and Desanti's articles (Part I, Book Two). (Note by F. LL.)

[25] Carleman: *Equations intégrales singulières à noyau réel et symétrique*, 1923.

[26] Perhaps a less powerful method at the present time than Carleman's analytic methods. (Cf. Julia: *Introduction mathématique aux théories quantiques*, 1936.)

increasingly refined physical experiments led to a new concept, and this concept found its essential tool in the formulated theory, ready for them just when they needed it. The efforts of those who were apparently cultivating the theory of sets for its own sake, and the efforts of those observing and experimenting in their laboratories, unknowingly converged in a common undertaking. The theories of linear integral equations and Hilbert space extend in very varied directions today. The future doubtless has rich developments in store for "functional analysis";[27] we have good reason to believe that these can come from that true philosophy of mathematics[28] which consists above all in "sensing deeply the intimate and continued relationship between the abstract and the concrete."

[27] Cf. Volterra and Pérès: *Théorie générale des fonctionnelles*, 1936.
[28] Leon Brunschvicg: *Les étapes de la philosophie mathématique*.

39

PHYSICS AND CHANCE:
HAS SCIENCE CHANGED ITS
MATHEMATICAL BASIS?

by Théo Kahan

Elysian Colloquy

Science will be deterministic or it will not be.
HENRI POINCARÉ

Things sometimes seem more reasonable than men.
FÉLIX KLEIN

Moreover, nature is always stronger than principles.
DAVID HUME

And (in this battle) the loser gains more than the winner.
B. SPINOZA

IT was a memorable day. Early in the morning the Chevalier de Méré, a wit and an inveterate gambler, somewhat in need of a shave after a night spent at dice, burst into Pascal's home and asked him point-blank: "In how many throws can you expect to ring the bell[1] with two dice?" This was the Chevalier's problem; it was also the origin of the calculus of probabilities. De Méré had no suspicion that he had just caused the birth of a new mathematical discipline. It was an audacious undertaking, trying to submit to analysis events which depend on chance and which, by this reasoning, seem ordained to remain outside the realm of calculation. Pascal succeeded in taming chance and in doing so earned incomparable glory.

---

[1] That is, get two sixes.

However, the explication of nature, under the powerful impetus of men like Galileo, Descartes and Newton, was taking a very different road. Ignoring chance and its intervention in natural events, scientists and mathematicians were committed to the construction of a deterministic dynamics, with the aim of gradually embracing all natural phenomena.

It was not until the second half of the 19th century, with the appearance of the kinetic theory of matter, that the prestigious work of Maxwell, Gibbs and Boltzmann opened the door to chance. But this chance, it must be said, was the chance of ignorance, a makeshift, which scientists were seeking to reduce to underlying dynamic laws, i.e., to a determinist scheme.

Everything went well until about 1900. With the appearance of *quanta* at the dawn of the century, serious doubts arose concerning the validity of using differential equations to describe natural phenomena. This happened to such a degree that in 1911 Poincaré uttered a veritable cry of distress:

> What these new investigations seem to be challenging is not only the fundamental principles of mechanics, but something that up to now has seemed inseparable from the very notion of natural law. Will we still be able to express these laws in the form of differential equations?

In other words, was physics going to change its mathematical basis?

The answer of modern science will be dialectic. One need only glance at the present situation in theoretical physics to be convinced of this. What, in fact, was the scientific ideal of classical science? It was hardly different from the Cartesian ideal: to describe reality by *extension* and *motion*. Science was going so far as to forget and lose sight of Pascal's warning:

> It must be said that by and large this is done by extension and motion, for that is true; but to say what these are and to construct the mechanism is ridiculous, for this is useless, uncertain and arduous.

The problem was to conceive an image of nature capable of being represented by schemes borrowed from the arsenal of mechanics. The relative motions of the various component parts performing their evolutions in space and time, and obeying the principle of universal determinism, were described with the aid of the differential equations of rational mechanics. The observer could disregard the fact of his

own presence, since he was not participating in any way in the unfolding of these phenomena.

Classical dynamics is thus deterministic, i.e., if one knows with
precision the state of a physical system,[2] one can calculate with equal
precision the subsequent value of each physical magnitude, such as
energy or impulse, associated with this system. However, Maxwell and
Boltzmann felt obliged to resort to statistical methods for certain
problems in the dynamic theory of gases. In fact, if one does not know
all the parameters but only certain of them, one can at least predict
the probable course of the system's development by averaging the
remaining unknown parameters. The same method would allow one
to indicate the probable development of other systems under investigation.

The dynamics of gases is a wonderful illustration of the state of
things. In order to describe with absolute rigor the varying states of a
gram of helium gas, one would have to know the positions and
velocities, at a given instant, of each of its atoms. In reality the
dynamics of gases calls for only two parameters, pressure and temperature, which are complicated functions of the innumerable
parameters hidden in this gram of helium. Thus this dynamics can
furnish nothing but probabilities.

The situation in wave mechanics is entirely different. In fact the
statistical predictions there present a radically different character.
The state of this system is described by a wave function Ψ which
satisfies a differential equation of the classic deterministic type.[3] If
one knows Ψ at the instant $t = 0$, its value at the instant t can be
calculated with precision. But this function, although completely
defining the physical system to which it belongs, furnishes only
probabilities. And so the indeterministic calculus of probabilities has
replaced deterministic dynamics until further notice.

If one tries to interpret the indeterministic character of the con-

---

[2] If n is the number of degrees of freedom of the system, this state is defined by
$2n$, given constants of which n are coordinates and n moments.

[3] Here is an equation for the simple case of the free motion of a corpuscle of
mass M and energy E along the x-axis:

$$\frac{d^2\Psi}{dx^2} + \frac{8\pi^2 m}{h^2} \times E \times \Psi = 0$$

where h is Planck's constant and $|\Psi|^2 \, dx$ measures the probability of finding the
corpuscle between the abscissas x and $x + dx$.

nections among the magnitudes by taking rational mechanics as a model, one could reason thus. In reality the wave function Ψ does not give a perfect description of the state of the system; for this knowledge to be precise, one must have still more givens at one's disposal. Therefore if pre-quantum physics is incorrect and incomplete at the atomic level, wave mechanics would be correct but *incomplete*; hence, the necessity for a theory yet to come that would be correct and *complete*. In other words, the atomic system would possess still other parameters and coordinates beyond Ψ. If we knew them in their entirety, we could predict the exact values of all physical magnitudes. By contrast, using only the function Ψ, we can predict nothing but probable values, just as in classical physics when only some of the parameters are known. This "deterministic" interpretation of wave mechanics is, of course, only hypothetical. The value of this speculation depends on whether we succeed in actually discovering these "hidden" parameters which do not figure in Ψ, and in building a deterministic theory with their aid. This theory would have to adhere to reality, and in addition, in the case where only the wave function is available, would have to reduce to the statistical predictions of wave mechanics by taking the averages of the other parameters).

It is incontestable that resorting to hidden parameters has already made it possible to reduce properties seemingly statistical in nature to deterministic mechanics. Think of the example we just gave, helium gas; the hidden parameters are precisely the coordinates and velocities of the individual atoms, and the pressure and temperature, *mutatis mutandis*, assume the role of the wave function Ψ.

Atomic entities can be grasped only by mental procedures.
F. ENGELS

Is such an interpretation of wave mechanics possible with the aid of hidden parameters? A very controversial and important question. Some of the greatest scientists—Einstein, Planck, Schrödinger and others—answer in the affirmative. If their thesis were to be substantiated, the current theory would be labeled provisional, and the description of states by means of the wave function would come to be regarded as intrinsically incomplete. But we shall see that the introduction of hidden parameters could not take place without singularly altering current theory.

Thus, numerous theoreticians have felt impelled to abandon the idea of hidden parameters. As a consequence of the work of Neumann, Solomon and others, they agree that the natural laws governing phenomena are statistical in nature. To give a specific example, let us suppose that we are playing heads and tails. The probability of getting a head on a single toss is 1/2; similarly, the probability of tails on a single throw is 1/2. However, if one performs many tosses, the number of cases in which one gets heads (or tails) divided by the total number of throws tends towards 1/2 and the discrepancy between the empirical frequency and the theoretical probability diminishes with the number of throws. If we accept the impossibility of predicting the results of a single toss—the fundamental indeterminism of a microscopic event —it is no less true that macroscopically (for a very large number of events) these events will be predictable, the precision increasing with the number of events. Determinism on the human and astronomical scale would therefore be due to the leveling effect of the "law of large numbers," which comes into play in the course of the simultaneous interaction of numerous elemental phenomena.[4] This law does not necessarily imply the determination of the individual event. What it postulates is that this determination obeys the calculus of probabilities. In other words, whereas the law is deterministic in nature, the individual phenomenon reveals itself to be random.

This point of view considers probability the real principle of natural law, and for this reason abandons the classic form of determinism; it goes back to Max Born, and constitutes the statistical interpretation of wave mechanics. According to this school, there is no direct evidence to support determinism, nor could there be because, without going into detail, the only reason for the apparent deterministic order of the world on the human scale is the law of large numbers. However, this law definitely does not settle the question of whether or not the natural laws governing atomic phenomena are deterministic in nature. Inasmuch as the number of atoms involved in the smallest experiments is immense (around 10^{25} in a gram of matter), the fluctuations are infinitesimal. Therefore determinism could only be put to the test in the atomic realm, at the level of elemental phenomena themselves. But in our present state of knowledge of the quantum, everything at this level seems to oppose determinism, for wave mechanics, the only theory which orders and unifies our experimental knowledge to a

[4] See the articles by Fortet and Servien in Part I, Book Two.

satisfactory degree, is in logical contradiction to determinism. (Von Neumann.)

Let us be specific. In an isolated microsystem, development proceeds according to a wave function Ψ governed by a differential equation of the deterministic dynamics type, but the act of observation performed by the experimenter substitutes a group of probability statements for a precise prediction. To get a better grasp of the nature of this investigative act, let us analyze the act of measurement more closely, along the lines of Heisenberg, Darwin, Weizsäcker and Neumann. Suppose we wish to determine the length of a rod.

1. We can observe the correspondence of this rod with a graduated ruler and say, the rod measures so many centimeters.

2. Or again, we can arrive at this result by calculation and obtain the length starting with temperature, etc.

3. We can also continue the calculations, deduce the behavior of the rod from its molecular properties and say, such and such a length will be observed.

4. Going further still, we could take into account the light source, evaluate the reflection of the photons by the rod as well as the path of the photons in the experimenter's eye before an image is created on his retina, and then say, such and such an image will be registered by the observer's retina. If our physiological knowledge were more precise than it is today, we could go even further by pursuing the chemical reactions which the image stimulates in the retina, optic nerve and brain, and finally conclude, such and such are the chemical changes perceived by the observer in his cerebral cells. But however far back we push the calculations, rod, retina, brain, at a certain moment there is a hiatus between subject and object; at that moment we are obliged to say, "Here is what the observer is perceiving." In other words, we always have to divide the world into two parts, one the observer, the other the observed system. In the latter, we are free to pursue all physical phenomena with greater and greater precision. In the other, the consciousness of the observer intervenes. On the atomic level, the notion of an object cannot be employed without taking the subjective act of knowing into account. The position we are developing is clearly different from idealism; the object is there, and no conjuring trick can make it disappear. It goes without saying that it is not the subject, with his emotional states and personal future, that is introduced in this way in physics. So then, two functions of consciousness are involved in every atomic description: knowledge and will. This

follows clearly from the fact that the wave function Ψ represents the probability relative to each *possible* result of each *possible* experiment. The first "possible" expresses our ignorance; a phenomenon is possible in this sense if we do not know whether or not it will occur. The second "possible," on the other hand, expresses a capability or an act of will; an experiment is possible in this sense if it can be carried out or omitted at will. There is therefore nothing objective about the caesura between the known and the unknown. It can be placed wherever one wishes, but it can never be done away with. Thus, in the example discussed above, there are several possible places for the hiatus; either the rod, the eyes or the optic nerve can be taken to be part of the observer. And this movable boundary between the observed and the observer inevitably imposes itself somewhere in the course of every description of an actual observation. In fact, experiments lead to statements of the type; "An observer has noted such and such an observation," and not to propositions of the type; "An atomic magnitude possesses such and such a definite value." And these perceived data always correspond to a macroscopic measurement of magnitude such as the position of the needles on a gauge, the lines and points on a photographic plate, etc. Between these perceptions and the atomic magnitudes there is a no man's land, which one gains possession of only with the aid of just that collection of mental steps which constitute a theory. Now, by means of a wave which behaves in accordance with a deterministic mathematical scheme, wave mechanics describes events occurring in the observable part of the universe, as long as these events do not interact with the observer part; but as soon as an interaction (observation) occurs, the deterministic mathematical scheme gives way to probabilities (reduction to wave packets).

Classical analysis, in the form of differential equations, adheres to the world which can be understood by the mind, the calculus of probabilities to the world which can be perceived by the senses.

> *There is an abundance of everything, except common sense.*
> ERNEST RENAN

Among the ancients, the atom, while lacking tangible qualities such as color, taste, etc., retained the quality of extension. In modern physics atoms lose even this last quality; extension is no more privileged to represent the atom than are the other qualities. The atom of quantum physics must resign itself to being symbolized by a partial differential equation governing the behavior of Brogliean waves in an abstract space of more than three dimensions. The physicist seems merely to be juggling formulas; the truth is that he is imprisoning the universe in them. From being a spectator, the observer becomes an actor. By the experiment which he carries out on the atom, he imposes on it a certain position, a certain color, a certain temperature. However one must not think that the basic interdetermination of the magnitudes of a state are due to the disturbances introduced in the object through the observer's act of intervention. This would leave one free to believe that the micro-object, even before it is observed, possesses certain properties which only the act of observation would alter. In the eyes of the quantum physicists this would bring us back to a pre-quantum mode of thinking. What is true is that in order to be able to attribute a definite property to an object, one must first have available a measuring device which permits us to establish this property. If by using another measuring device, one proceeds to measure a magnitude that complements the magnitude measured with the first device, then the very conditions are destroyed under which the magnitude measured in the first place could have a definite value. It is this physical intervention, necessary for proceeding from the old experimental conditions to the new, that is commonly called "perturbation of the object by observation." This expression has a precise meaning only if one subjects, not the fictitious, non-perturbed object, but an object already known by observation, to a new observation of another type.

   The physics of indeterminism is just now celebrating its 20th anniversary. Can it be said, as the quantum physicists maintain, that determinism is definitely finished? As to determinism, we are certainly dealing with a deeply rooted mode of thought, but certainly not with an intellectual imperative, as the very existence of a doctrine such as quantum mechanics proves to be.
   Thus in the probabilistic interpretation, at the atomic level, the unit of intelligible reality is no longer the isolated object, but rather the man-thing relationship as a whole. It cannot be emphasized too much that the atom is not directly perceivable. It is not given as an object in space and time, but as a term in an inference derived from an

experiment, thanks to certain procedures of thought. Nor can it be described by means of a model patterned on a macroscopic object possessing extension at a given moment. On the contrary, it is characterized rather by a certain mathematical magnitude, the Brogliean wave function Ψ. This function takes the place of the mechanical magnitudes which we ordinarily use to describe an object and which permit us to predict its future behavior. How does this wave function accomplish its mission? An experiment always results in a definite magnitude in space and time; since the function Ψ is not such a magnitude, it is not directly accessible (measurable). It furnishes us only with probabilities about the results of the experiment.

Is the probabilistic interpretation, as well as the corresponding change in our concept of an object, definitive and inevitable? What do we mean however, when we speak of the definitive character of a physical theory? A coherent theory entails rigorous conclusions. Now, it is impossible to prove with complete rigor that a theory "holds" perfectly. Science has therefore felt obliged to associate with every physical theory the idea of a corresponding domain of validity. Thus a theory will be considered correct, not when it embraces all conceivable facts, but when it correctly represents a group of reproducible facts. If another theory embraces a larger group of phenomena, the first theory will be a special case of the second. Thus classical physics possesses a certain domain of validity, wave mechanics another, more extensive one, each theory preserving its own character definitive over its respective domain of validity.

Consequently the question concerning the definitive character of quantum mechanics takes the following form: Will not somebody construct in the future a theory more complete than wave mechanics, which will permit us to obtain objective values for magnitudes now reputed to be indeterminate? This possibility is not logically excluded. Only, a limiting condition will have to be imposed on the new theory; in fact, it can be seen that the theory will be obliged to forgo describing phenomena perceived by means of classical physics, since the latter describes phenomena that are perceivable on the human scale by employing the concepts of corpuscles and waves in the framework of space and time. This new and completely hypothetical theory would therefore consider itself obliged to resort to a different description of natural phenomena, so new and radical in style that the classical distinction between corpuscles and waves would be wiped out. The behavior of a localized corpuscle would then depend upon the state of the entire space in a way that would, from the classical point of view, denote the presence of an extended field. Because of the difficulty of sharing Schrödinger's conception of such a state of things, as well as the fact that various efforts in this direction have run aground, quantum physicists now seem inclined not to bank on a revision of this type in the guiding principles of contemporary science.

How is it that at present we are unable to produce an object in isolation? We cannot overemphasize the fact that if our knowledge rests on experiments, these constitute consciously performed acts of intervention. Now when we intervene it is by means of a device manufactured by us. The result of this intervention is a perception. Therefore the physical progress of an experiment must be described within the framework of the space and time of our intuition. Moreover, a strict determinism must govern the functioning of our measuring device; otherwise we could not draw unambiguous conclusions from the result of observations on the observed object. In addition, we must think of the objects making up the experimental apparatus as material bodies; as a matter of fact these remain identical with themselves just as material bodies of daily life do. This is indispensable if we wish to set forth in tangible terms what we have measured. Thus our description of the experimental apparatus uses as a frame the concept of space-time as well as the categories of substances and of determinism. What we actually observe are the needles of a gauge, the traces left on a photographic plate, etc. Between the atom and the observation is a sort of *vacuum* which the theory must fill. The atom is therefore comprehensible only by virtue of these mental processes.

If one could predict with certainty the results of all the experiments conducted on the atom, simultaneous experiments relating for example to its position, velocity and internal structure, then thanks to these fixed attributes, one would be able to attach to the atom the values of magnitudes—all defined in terms of space-time—which were obtained through these measurements. We would thus obtain a space-time model of the atom. Now, precisely the fact that statistical statements exist shows the present impossibility of attributing measured magnitudes to the atom as if they were characteristic properties. What is more serious, measured magnitudes conceived as attributes of one and the same object are contradictory. Thus, the helium atom, for example, in the course of certain experiments behaves like a spatially concentrated corpuscle, and in the course of other experiments like a

Brogliean wave filling space. Now the helium atom cannot be wave and corpuscle at one and the same time. The logical contradiction is resolved by the fact that one can never simultaneously carry out experiments in the course of which the atom reacts in two different ways. If one has marked the location of an atom at a specific point in space, an immediate repetition of the experiment will find it at the same point, and one will properly be able to say, the atom is in such and such a place. In this case, the atom's function Ψ is such that it can furnish only probabilities for the other, complementary magnitudes. Conversely, if one knows its wave properties, one cannot predict its corpuscular attributes except with a certain probability. One no longer has the right to say that the atom is a corpuscle, or the atom is a wave, but that it can be either wave or corpuscle. In short, the experimental device decides the form in which the atom will appear. Man *forms* the object in the process of inquiring about it. Are there good grounds then for Poincaré's concern? Shall we still be able to express the laws of physics by means of differential equations of classical analysis?

To use Denjoy's language, if the theory of functions suggests that matter be considered as a perfect and totally discontinuous set, then this set, depicted on our ordinary level, is equivalent to a finite number of separate and distinct blocks. But when this process of division is pushed indefinitely, the searcher's gaze falls upon a singular archipelago formed of sharp coral reefs. He would try in vain to anchor his imagination there, for this indescribable landscape would be rocked by who knows what tenuous and uncertain tremors brought on by the shock of his very gaze and by the mere process of his perturbing investigation. This representation of matter, subjected to an indefinite analytical development (in the form of solutions of differential equations), leads us therefore to a terminus where every analogy with nature, perceived in its simplicity, and on our scale, has disappeared. Analysis applies to undisturbed nature; its reign ends where man and nature meet, and there it gives way to the calculus of probabilities. Here then is the answer that modern science brings to Poincaré's anguished question, a completely dialectic response since it is put on the level of a dialogue between man and nature.

Does this then make reality dependent upon our whims? Certainly not reality itself, but its image, the representation we use to understand it, is. Experimentation is an act of violent intervention in the course of events. It forces the atom to reveal its attributes to a mind incapable of conceiving events except in space-time. From this arises the impossibility of objectifying the various constituent elements of the atomic model, of joining and fusing the complementary aspects of the various sensible attributes of the atom, into a kind of unity independent of the comprehending subject. All the qualities of the corpuscle of theoretical physics are derived qualities; it does not directly possess, independently of the experiment which the observer carries out on it, any primary material attribute, in the sense that any kind of image we might care to employ to concretize the atom is *eo ipso* tainted with error. An immediate comprehension, intuitive in space-time-energy terms, is consequently impossible in the universe of particles. Sensible reality transcends the framework of space-time where, it was thought, one could perceive and follow the course of events. In making nature transparent, we act upon it. Man's technical activity, "praxis," transforms the universe as it probes it for knowledge.

Thus the a priori category of Kantian determinism is fought to the breach as were the a priori forms of sensibility before it. Shed of all its a priori tinsel, the ontological principle still stands. The universe *exists*, coherent and rational. When we explore it, reason and reality mix intimately. The dialectic of knowing is substituted for the dialectic of being. If the general law remains rigorously deterministic, the particular event is essentially random.

Is it therefore certain that physics did the right thing when, by going from the certain to the random, it changed its mathematical basis? Are we then forever barred from finding an underlying determinism beneath the indeterminism of the quantum? Let us recapitulate. At the atomic level classical physics is incorrect and incomplete in the unanimous judgment of physicists; quantum physics is correct but *incomplete* (as a consequence of the hidden parameters) according to the determinists, but correct and *complete* according to the quantum physicists. What can be said with complete rigor is that it is impossible to complete quantum mechanics in the determinist sense without totally rejecting it, with its marvelous network of abstract, yet brilliantly verified, laws.

However that may be, any progress made in the matter of cognition alters our very idea of what "knowing" is. This, perhaps, is the reason for the difficulty we experience in specifying and defining "objective reality," which, from the viewpoint of the theory of knowledge, is at once so subtle and so solid. The ideal of knowledge changes with time. The idea of objectivity is itself dialectic, i.e., a function of a time and place in history. It does not admit of an eternal definition. Being quasi-autonomous in its development, it is free to find necessary, therefore causal, chains beneath historical accidents. It is no longer the object in isolation, but only the dialectical synthesis of the relation of knowing subject to knowable object, which quantum physicists consider the graspable reality at the heart of things. The atom is the material support of a mathematical scheme of proven relationships born through contact with experience and backed up by it; at the same time it is the center of convergence of all the experimental and intellectual operations which serve to define it.

Moreover, what is there to be gained by reducing the apparent indetermination of atomic phenomena to an underlying determinism, and what price would we pay in the form of complications? The progress achieved by quanta has led physicists to revise their notion of their own ideal. They no longer deem it their function to interpret nature with the aid of more or less anthropomorphic images, but only to give as good an explanation as the structure of their minds permits. Instead of aiming for an ontological, and therefore anti-dialectical, atomic essence, they intend to grasp what is quantitatively intelligible in reality, and this is mathematical in character. The scientist masters the prodigious complexities of reality infinitely better by means of formulas and abstract theories than by recourse to pictorial representations. Poincaré has said:

> If the Greeks triumphed over the barbarians and if Europe, the heir of Greek thought, dominates the world, it is because the savages loved their gaudy colors and the noisy sound of their drums, which occupied only their senses, while the Greeks loved the intellectual beauty hidden beneath the beauty of the senses and this is what makes the intelligence strong and sure.

By enfolding reality in a web of functional relationships, reason incorporates it into itself, creates new freedom of thought, and in the same stroke shows how the very foundations of our thought can be extraordinarily enlarged without the slightest concession to imprecision or incoherence.

The mind gets its impulse from the impact of reality and plunges into the ocean of things in order to fish up a miraculous haul of methods and hypotheses, of laws and theories, which it immediately hastens to confront with experience. And thus the cycle is completed. Thought returns, enlightened and enriched, to the experiment which called it forth, then sets out once more to conquer a nature which is at first rebellious and reticent, but which in the last analysis submits to its conqueror.

To apprehend the real world is to deal mathematically with the real world.

Book Four

ART AND ESTHETICS

MATHEMATICS AND BEAUTY

We have consciously distorted the portrait of mathematics and have rendered less than a full account of its relations with science. The same reason has led us, in the opposite direction, to multiply the evidence of bonds uniting mathematics and art. We are involved here with an area that has been rather poorly explored up to now, and where it will be worth our while to concentrate our efforts.

We have already noted that there are two currents which serve as a communications link between mathematics and the other human activities, one ascending, the other descending. The arts and beauty are no exception to this rule. There is a beauty in mathematics. This is not to be confused with the use of mathematics to create beauty in works of art. The esthetics of mathematics must be distinguished from the application of mathematics to esthetics.

It is to the first current, the one which attests to the beauty in mathematics, that we have devoted our study. We are anxious to refute one of the most persistent and widely held prejudices, one unjustly harmful to the reputation of mathematics, which the general public and especially those in the arts and letters are guilty of. A perusal of the other articles of this work will add an eloquent commentary to the thesis we are defending.

Adolphe Buhl's article includes metaphysical and moral considerations which would have justified classifying it in the chapter on philosophy. His especially lively appreciation of the beauty of mathematics made us decide to put it with the other articles on esthetics. The same characteristics are to be found in Buhl's personal work, particularly his work on Stokes's formulas.

"Number comes to life in Art," wrote Saint Augustine. And Leibniz went even further: "Art is the highest expression of an interior and unconscious arithmetic." In our opinion it is an excessive restriction on the nature of art to claim that it is reducible to mathematics. Nonetheless this is a penetrating view and a useful reaction against the platitude which places art and mathematics at opposite poles, the spirit of refinement and grace as against the spirit of geometry.

It would be a mistake, moreover, to limit the involvement of mathematics in art to numbers. Other notions independent of number, particularly the notion of group, *play a role there, as we are beginning to realize. Andreas Speiser is one of the finest experts today in group theory. The work which he published on this subject in Switzerland—he teaches at the University of Basle—has become a classic of its kind. We have in Speiser a mathematician who is also a humanist and a student of esthetics. If he has been able to trace the presence of mathematics in the fine arts so successfully, it is doubtless because he personally invests his mathematical works and expositions with so much art and elegance. His article, which breaks new ground in a still virgin territory, holds many surprises and a rich harvest for us.*

Speiser's article covers the entire domain of the arts, showing equal interest in the plastic arts, music and finally poetry. The articles of Le Corbusier and Henri Martin are focused more narrowly, the former on architecture and the second on music.

The text of Le Corbusier, who lays no claim to expertise in the domain of higher mathematics, is all the more interesting in that he speaks from his own actual experiences. This direct testimony from one of the most illustrious architects of our time confirms that it is not only through technique but also through sensibility that mathematics can enter into a work of art.

The problem of the relationships between mathematics and music is complex. Beauty here sometimes seems to be related to numbers and sometimes to groups (through the mediation of geometric forms, of which musical scores give a distorted representation). Musical beauty can be broken down into the various elements of a work: pitch, melody, harmony, tonality, modes, chord progressions, counterpoint, rhythm, musical form (fugue, sonata, etc.). Henri Martin's well-organized and clearly written article contains the essentials of what one should know about the Western musical scale: This numerical basis of music governs several of its other aspects.

F. LL.

40

BEAUTY IN MATHEMATICS*

by François Le Lionnais

Circe never had as much power over her Ulysses as this marvelous science has over the mind, once its first difficulties have been surmounted.

R. BAUDEMONT

Be on guard against the enchantments and diabolical attractions of geometry.

FÉNELON

Mathematics, rightly understood, possesses not only truth, but supreme beauty.

BERTRAND RUSSELL

BEAUTY often appears at feasts where only utility or truth have been invited. How then can one remain insensitive to the fascinating charms with which she adorns them? This is true of all activities and all branches of knowledge, but nowhere with more force than in mathematics. Has not the modern Western world confirmed the opinion of ancient Greece, which up to the time of Euclid considered mathematics more art than science? Everything considered, it is most often alluring esthetic satisfactions which have motivated modern mathematicians to cultivate their cherished study with such ardor.[1]

Some of the most cultivated writers have attested to this fascination.

---

* We have put all the notes for this article at the end since we do not wish to interrupt the continuity and they are not indispensable in a first reading. Their purpose is to support our arguments or to give specific data on allusions made in the text.

As for the illustrations accompanying the text, we only expect them to enliven it, not to represent true beauty in mathematics. "What is most interesting in this country is not seen." (Henri Michaux: *Au pays de la magie*.)

Thus Novalis: "The true mathematician is enthusiastic per se. Without enthusiasm there is no mathematics." And, "Algebra is poetry."

Or the Goncourts: ". . . mathematics, and its enthralling power."

But it is the mathematicians themselves who have left the most passionate testimony.

Charles Meray writes:

> When we read Gauss's memoirs, whose fresh bloom is still unwithered after nearly a century, do not the details bring to mind those splendid intertwining arabesques conceived by the inexhaustible imagination of artists of the Orient? Does not the overall structure at the same time recall one of those marvelous temples which the architects of Pericles raised to the Hellenic divinities?

See how Painlevé recalls the teaching of Charles Hermite:

> Those who have had the good fortune to be students of the great mathematician cannot forget the almost religious accent of his teaching, the shudder of beauty or mystery that he sent through his audience, at some admirable discovery or before the unknown.

The eminent logician Bertrand Russell discerned perfectly this superior quality by virtue of which the queen of the sciences can lay claim to the crown reserved for the arts.

> The true spirit of delight, the exaltation, the sense of being more than man, which is the touchstone of the highest excellence, is to be found in mathematics as surely as in poetry.

If some great mathematicians have known how to give lyrical expression to their enthusiasm for the beauty of their science, nobody has suggested examining it as if it were the object of an art—mathematical art—and consequently the subject of a theory of esthetics, the esthetics of mathematics. The study which follows has no intention of establishing the latter; it aspires only to prepare the way for it. The materials that we are going to review will permit us to set up some rough classifications and will merely suggest a provisional basis for more penetrating studies.

Two criteria have guided us in clearing the ground. The first relates to the structure, not of mathematics itself, but of mathematicians' works. The second relates to man's conceptions of beauty.

Without in any way prejudging the real nature of mathematics, one can assume it provides us sometimes with *facts* and sometimes with *methods.* This seems to us a useful distinction and sufficiently justified for an inquiry that is strictly esthetic.

Works of art, too, can be ranged under two grand banners—we do not say which of these we personally prefer: *classicism,* all elegant sobriety, and *romanticism,* delighting in striking effects and aspiring to passion. The two following passages indicate rather well the contrast between these two tendencies:

> I was struck by the art with which mathematicians remove, reject, and little by little eliminate everything that is not necessary for expressing the absolute with the least possible number of terms, while preserving in the arrangement of these terms a discrimination, a parallelism, a symmetry which seems to be the visible elegance and beauty of an eternal idea. (Edgar Quinet)

> What strikes us first of all, when we compare the mathematics of our times with that of previous epochs, is the extraordinary diversity and the unexpectedness and circuitousness of the paths it has taken; the apparent disorder with which it executes its marches and counter-marches; its maneuvers and constant changes of front. (Pierre Boutroux)

Classicism	Romanticism
There, all is order and beauty, luxury, calmness and delight.	*There are lines which are monsters.*
BAUDELAIRE	DELACROIX

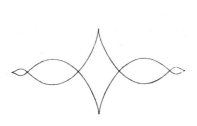

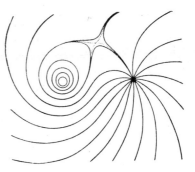

Curve obtained by applying a Joukowsky transformation to an asteroid (= hypocycloid with four cusps).

Some curves satisfying the differential equation:

$$y' = \frac{x[(x-a)^2 + y^2] - y(x^2 + y^2)}{(x-a)(x^2 + y^2) - y[(x-a)^2 + y^2]}$$

Thus armed with this double-edged blade, we are going to develop our idea with copious examples, each of which will give us an opportunity to provide details we consider important. After considerable hesitation we have chosen to gather this sheaf of examples from the field of elementary mathematics. By so doing, we have been obliged to pass over some of the purest and most vivid examples of beauty in mathematics, in the hope of reaching a greater number of readers. As Carl Stoermer wrote:

> What one learns about mathematics in primary school corresponds to the alphabet. What one learns in high school corresponds to the sentences of a primer. What one learns in elementary college courses corresponds to simple little stories. Scholars alone are aware of the mathematics that corresponds to literature.

Classical Beauty in Mathematical Facts

We say that a mathematical proposition has classic beauty when we are impressed by its austerity or its mastery over diversity, and even more so when it combines these two characteristics in a harmoniously arranged structure.

Regularity strikes and intrigues us especially when we are expecting a certain disorder. This bounty is all the more delightful in that to some it seems unmerited, to others won after a mightly struggle.

Who does not know magic squares? These are cross-ruled squares, filled in as if at random with numbers belonging to a set of consecutive whole numbers. Just as we were resigning ourselves to chaos, we notice the unvarying repetition of a single total for the main directions of this grill. Since one does not need to be endowed with mathematical knowledge to enjoy this pastime, we can understand its having diverted, and sometimes confounded, so many good people all the way back to Renaissance times. We recall the magic square[2] which, like an arithmetic marquetry, Albrecht Dürer introduced into his "Melancholy" (p. 125), in addition to numerous allusions to geometry and mechanics.

One of the most modest branches of the tree of mathematics, the geometry of the triangle, has always had its faithful admirers, because of the graceful ornaments which adorn it. Is it not wonderful to note, throughout this brief episode in the great story of geometry, how often three straight lines meet in a single point, or three points lie on one straight line? I have sometimes thought that it would be worthwhile to

Albrecht Dürer—"Melancholy"

At this period I would have been less inclined to see unknown kingdoms than to be acquainted with theories.

ALBRECHT DÜRER

begin by offering students pictures wherein the three altitudes of a triangle do not meet (misleading them with an incorrect construction), before teaching them that it is impossible for the lines not to meet.

It is equally surprising to discover more than three points (especially if they come from different definitions) appearing on the same circumference, since three points are sufficient to define a circumference and a fourth is not likely to be on it. In any given triangle the centers of the three sides, the feet of the three altitudes and the centers of the three segments joining the orthocenter (the point in which the altitudes meet) to the three vertices are situated on one and the same circumference, called the *nine-point circle*, or preferably, *Euler's circle*. Thus nine points, furnished by three different definitions, come together on the same circumference, like ballet dancers in a choreographic figure. The acrobatic genius of Euler surely savored this decorative miracle. How he would have marveled had he known how many stars would be added to his original *corps de ballet* in two centuries. There are now 31, and perhaps 43, different points blooming on this mathematical garland.[3]

The study of curves in classical and analytic geometry is equally blessed with harmony. Has not the cycloid,[4] found in so many natural phenomena, been called the "Helen of geometry"?

The logarithmic spiral,[5] discovered by Descartes and studied chiefly by Jacques Bernoulli in a treatise on the differential calculus published in 1698, possesses numerous astonishing properties, notably that of being equal to its caustics by reflection and refraction, to its evolute, and to numerous other derived or conjugate curves. For this reason Bernoulli requested that on his tombstone be engraved a logarithmic spiral, above the following inscription: *eadem numero mutata resurgo*. "This marvelous spiral," he wrote, "gives me such overwhelming pleasure by virtue of its singular and wonderful properties that I can scarcely satisfy my desire to contemplate it."

Similar bursts of fireworks shoot forth and disperse, as if spontaneously, at each step of the theory of functions. Who has not been amazed to learn that the function $y = e^x$, like a phoenix rising again from its own ashes, is its own derivative?

We appreciate that these various examples are very unequal in value. The value of each of them depends on the depth of the mathematics required to prove it. Let us recollect that what we are studying here is the subjective character of the beauty in mathematics; we are not concerned with the objectivity of the mathematics itself. A fact which overwhelms us the first time we meet it, comes to appear trite in the end. Thus the mysterious novelty of the preceding example is dissipated to a large extent as soon as we realize that it is not very surprising for the differential equation $y = y'$ to have a solution. Still a certain margin of wonderment remains when one realizes the economy of this solution.

Imagine my surprise (I was going to say, my disbelief) when I learned as a schoolboy that the product of the least common multiple of two numbers and their largest common divisor is exactly equal to the product of these two numbers. My wonderment completely disappeared when I realized a little later that this property obviously results from the definition of the least common multiple and the largest common divisor based on the composition in terms of the prime factors of the initial numbers. Awareness of a mechanism dethroned the impression of finality and extinguished its iridescence.

This impression of finality often plays a large role in the esthetic enjoyment which the sciences can provide us. There is, to be sure, no finality in mathematics, any more than there is in nature, but the very frailty of our intelligence engenders these illusions which so stir our emotions.

Without as yet leaving the empire of classical beauty, we can add the delightful Ionian slenderness to the rigid simplicity of the Doric order. But at this stage of our inquiry, we shall invoke diversity only when it is not excessive and when monotony seems to threaten.

This is what happens in plane trigonometry when one runs into the formulas for the addition of arcs; the rhythmic complexity of their structure agreeably shatters the anticipated poverty.[6] And what shall we say of the celebrated hypergeometric series, this Proteus of mathematics, whose ability to metamorphose itself into highly varied functions by means of trifling modifications of its coefficients is veritably unbelievable?

It is always impressive to emerge from a long, wearisome trek underground into the blue open, among the high peaks. The view encompasses horizons revealing unsuspected surroundings. Is it such a sentiment which inspired Gauss to describe his celebrated theorem "on quadratic reciprocity," relating to prime numbers, with the phrase "the jewel of arithmetic"? What is more startling than this theory, on whose reefs the efforts of the cleverest mathematicians have gone aground? Every prime number seems immobilized by a tight girdle of steel which prevents all communication with the other numbers. We must keep this rigorous restriction in mind if we wish to appreciate fully the importance of Gauss's discovery, which permits two prime numbers to exchange roles, for all the world like two trapeze artists crossing in mid-air.[7]

Pascal's arithmetic triangle, also, which is very easily constructed,[8] provides a most remarkable link to the coefficients of Newton's binomial expansion,[9] to the important notions of "combinations" and "permutations" encountered in combinatorial analysis,[10] and consequently to the calculus of probability.

When the same mathematical notion can be defined in two different ways, this very fact introduces a correlation between the subjects of these varying definitions that often would not have been suspected. This is notably the case with the number π whose geometric definition everyone knows and which can be equally well expressed by formulas from analysis, or by very different series of numbers.[11]

The same is true of the number e. Sometimes its definition is encountered at the beginning of the study of the theory of functions, sometimes at the beginning of the study of logarithms. Like the little phrase in Vinteuil's sonata, it later reappears in unexpected series.[12]

After numerical series, there are functional series. What student has not been dazzled upon first meeting the Taylor and Maclaurin series,[13] whose gold and silver chains link together the sparkling gems of the successive derivatives of the same function, on up to infinity?

Euler's formula $\sqrt{-1}^{\sqrt{-1}} = e^{-\pi/2}$ (which can also be written $e^{\pi\sqrt{-1}} - 1 = 0$), establishes what appeared in its time to be a fantastic connection between the most important numbers in mathematics, 1, π, and e. It was generally considered "the most beautiful formula of mathematics." The brilliance of this expression is due to the nearly perfect elimination of every element foreign to the three numbers just cited. Today the intrinsic reason for this compatibility has become so obvious that the same formula now seems, if not insipid, at least entirely natural.

Sometimes rather simple transformations can alter one of a pair of traditional curves into the other though they had appeared quite unrelated. The sudden revelation of their kinship is truly a treat.[14]

There are relationships even more monumental, such as that bridging the gap between algebra and geometry: this striking dualism, which associates a figure with every equation and vice versa, each of them holding a mirror wherein the other is reflected, is one of the most

On a strange whim I had banished irregular plant forms from these scenes.
                              CHARLES BAUDELAIRE

It was an endless palace full of pools and waterfalls tumbling into dull and burnished gold.
                              CHARLES BAUDELAIRE

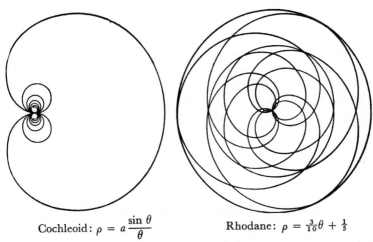

Cochleoid: $\rho = a\dfrac{\sin\theta}{\theta}$

Rhodane: $\rho = \frac{3}{10}\theta + \frac{1}{5}$

Her caresses were so light that pathways sprang up by themselves at each instant.
                              PAUL ELUARD

... the Wonderland so sweetly dreamed of.
                              LEWIS CARROLL

Differential equation:

$$y' = \frac{(y\tan\frac{\pi}{6} - x + a)(x^2 + y^2) + \left(x - y\tan\frac{\pi}{6}\right)[(x-a)^2 + y^2]}{[(x-a)\tan\frac{\pi}{6} + y](x^2 + y^2) - \left(y + x\tan\frac{\pi}{6}\right)[(x-a)^2 + y^2]}$$

Differential equation:

$$\rho\,\frac{d\theta}{d\rho} = \tan\left(4\arctan\sin 4\theta + \frac{\pi}{4}\right)$$

distinctive in all of mathematics; truth, utility and beauty, joined in intimate marriage, give birth to the most glorious and vital perfection.

Next in turn, infinitesimal analysis reveals to us the sublime interdependence between the area and tangent of a curve. The relationship which unites the curve and its tangent at a given point is expressed analytically by the derivative[15] of the function which describes the curve. Moreover, the relationship which unites the curve and the area bounded by it is expressed analytically by the integral[15] of this function. Now, by definition, integration is nothing but the reverse of differentiation. The result is that the tangent to a curve and its area are only inverse modifications of each other with respect to the law describing the inherent structure of the curve itself.

Romantic Beauty in Mathematical Facts

By contrast with classic mathematical beauty we are now going to examine another sort of beauty which can be described as romantic. Its underlying principle is the glorification of violent emotion, nonconformism and eccentricity.

The notion of the asymptote, with which we shall open this new series of examples, has as it were fallen into the public domain. Thus it requires an effort to appreciate to what extent an entire epoch could have been intrigued by its discovery. Montaigne writes,

> Jacques Peletier was telling me at my house that he had found two lines approaching each other, which, however, he established could never succeed in meeting except at infinity.

One must guard against automatic judgments; more conscientious investigation will sometimes bring unexpected confutations. Ask a high school student what would be the result of raising a given number to the zero power. He will respond with assurance that the result is zero, but then discover to his shame that it is 1, no matter what number the base is. One experiences the same kind of surprise— as Silvanus P. Thompson notes—upon realizing

> how, often, there is little resemblance between a differential equation and its solution. Who would suppose that an expression as simple as
>
> $$\frac{dy}{dx} = \frac{1}{a^2 - x^2}$$

Luxury, oh hall of ebony, where garlands of renown writhe in their death throes to beguile a king.
STÉPHANE MALLARMÉ

From monster to monster, from caterpillars to giant larvae, I went, clutching my way.
HENRI MICHAUX

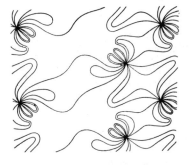

Pascal's limaçon transformed by elliptic functions

. . . worlds of a sardonic reality brushing against whirlpools of feverish nightmares.
HOWARD PHILLIPS LOVECRAFT

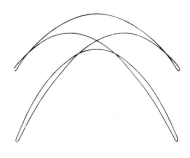

Algebraic equation:

$$20y = (8 + x \pm \sqrt{16 - x}) \times (8 - x) \pm \sqrt{16 + x}$$

. . . algebra danced madly.
ALDOUS HUXLEY

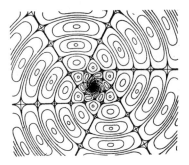

Simultaneous differential equations:

$$\frac{du}{dt} = \frac{\tan u}{\tan t};$$

$$u = \theta - \rho; \quad t = \theta - \frac{1}{\rho}$$

Simultaneous differential equations:

$$\frac{du}{dt} = \tan 2u; \quad u = \sin \rho - \theta;$$

$$t = \int \frac{d\rho}{\rho^2 \cos \rho} + \theta$$

could be transformed into

$$\frac{1}{2a} \log e \frac{a + x}{a - x} + C \quad ?$$

This resembles the transformation of a chrysalis into a butterfly!

Amazement can turn into something still more violent, producing what seem to be completely illogical results repugnant to common sense.

One of the most difficult branches of mathematics, *analysis situs*, that spring whose waters lose themselves in the ocean of modern topology, abounds in spells and charms. Certain of these, like the one-sided Möbius strip, are as amusing to children as conjuring tricks,[16] even though arising from serious problems.

One always experiences some difficulty in getting those uninitiated in the calculus to appreciate the mechanical properties possessed by the cycloid, whose classic beauty was mentioned above. These properties were discovered by Jacques and Jean Bernoulli and Christian Huygens; some malice must be concealed in the barbarous names given these properties, brachistochronism and tautochronism.[17] The a priori conceptions of common sense burst into pieces under the hammer blows of mathematical analysis, and in a harsh light their fragility becomes apparent.

The propensity of the mathematical spirit for escaping physical reality on the wings of its rational imagination often provokes it to fashion concepts in which the uninitiated are prone to see insane nightmares rather than the fruits of logical activity. Of course, in the end one grows accustomed to anything; in short, it is their success which sanctions these innovations, at the same time that it robs them of their charm.

To someone who knows how to raise a number to the 2nd, 3rd, 4th, etc., power, i.e., multiply it 1, 2, 3, etc., times by itself, what indeed would raising it to a fractional or negative power, like 3/5 or −4, correspond to? To nothing, evidently, as long as we have not agreed to extend the original definition in such a way as to permit its adaptation to new situations. But what could be said about a power whose exponent is imaginary, i.e., contains $\sqrt{-1}$? How could one hope to exorcise such a phantasm? Happily this is possible and by no means involves a gratuitous rule of the game. It is indispensable in obtaining the previously mentioned Euler's formula, which plays an essential role not only in pure mathematics but also in the application of science to technology.

At the beginning of the 19th century, Poncelet was led by the study of the intersections of two ellipses to introduce the notions of *circular points* and *isotropic straight lines*. These last enjoy extraordinary properties which cause students much delight: any two given circles intersect in the circular points; every isotropic straight line forms any angle with itself and in particular, a right angle—the last property being frequently used; the distance between any two points on this straight line is always zero, etc. We are not dealing here with jokes in questionable taste. Mathematics remains perfectly serious even when it affects these lunatic airs.

For centuries the impossibility of proving Euclid's parallel postulate[18] was "the scandal of geometry and the despair of geometers" (d'Alembert). Its replacement by other postulates, those of Lobachevsky-Bolyai[19] or those of Riemann,[20] seemed an intolerable *coup d'état*. And yet without this replacement could we have developed the general theory of relativity, which plays so powerful a role in explaining the universe?

Many people would have bet against the appearance of the number π in the calculus of probabilities. True, a little reflection will destroy this impression if one is familiar with the experiment known as Buffon's needle.[21] In fact, one finally realizes that the figure needed to illustrate this experiment has something to do with circumferences and their diameter. To penetrate the enigma of how the number π enters into the probability that two integers picked at random will be prime with respect to each other,[22] a knowledge not of elementary geometry, but of advanced analysis, becomes indispensable.

Does not the modern theory of sets take as its point of departure conceptions which seemed an insolent defiance of common sense when Cantor defended them?[23] This exuberant theory had to enjoy repeated successes in other disciplines already classic like arithmetic and analysis before we would accept the existence of quantities "greater than infinity" (Cantor's expression) plus that startling number ω situated *on the other side of infinity*. Theologians were not the last to protest certain ideas as unfair competition.

After the paradoxes come the anomalies, the irregularities, indeed the monstrosities. They arouse some people's indignation and to others bring delight.

The extension of the notion of multiplication introduced a whole series of related notion in mathematics that for convenience we still call multiplication; but these sometimes have the annoying property of no longer being commutative, with the result that the product of two factors is no longer the same but depends on the order in which they are multiplied. What perversity indeed![24]

Certain transpositions do not appear to be necessarily introducing anything unexpected. However we must not rush into stating this as a certainty before carefully verifying it. Are not the essentials of the straight line and the circumference contained in the respective properties of shortest distance between two points and equal distance from a central point? Look at the ravages produced when they are uprooted from the plane and transplanted onto the pseudosphere;[25] the straight line twists into an elegant loop; the circle penetrates its own interior, then breaks apart in hysterical contortions, and finally explodes into a spiral. And lastly the triangles, those humble assemblages of three straight line-segments, metamorphose into alarming hydras, the sum of whose angles is always less than 180° and may even become zero.[26]

The following example is no less subversive. It concerns the generalization of the notion of circle in the geometry of n dimensions, n being equal successively to 2, 3, 4, 5, etc. The definition of such a figure always remains the same: A "hypercircle" in a space of n dimensions will be formed by the figure bounded by all the points situated at an equal distance from an interior point called the center. For each of these successive spaces, it is easy to calculate the formula for the measurement of a hypercircle in a space of any given number of dimensions.[27] For simplicity, let us suppose that the length of the radius is always equal to unity. It is disconcerting to note that the measurement of the hypercircle first keeps increasing, reaches a maximum, and then constantly decreases and approaches zero. The most disturbing aspect of this business is that the maximum is located in a space the number of whose dimensions is not an integer and lies between 7 and 8.

When Riemann and Weierstrass made known the existence of continuous functions without derivatives, what an outcry came from the mathematicians against these newcomers. "I turn with fright and horror from this lamentable plague of continuous functions having no derivatives," exclaimed Charles Hermite. If it is difficult to reason about such functions, it becomes impossible to visualize fully the infinite caprices of the curves representing them. The arcs joining two

Yes, these are reflections, negative images
Tossing themselves about like a motionless object
Throwing their active multitude into the nothingness
And composing a counterpart for every truth.
                                    RAYMOND QUENEAU

HERE IS WHAT BECOMES OF THE FOLLOWING
WHEN THEY ARE EXILED TO A PSEUDOSPHERE:

Straight line	Circle (first kind)	Circle (second kind)
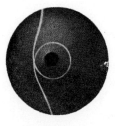		
Circle (third kind)	Circle (special third kind)	Any triangle

Triangle (with one zero angle)	Triangle (with two zero angles)	(Triangle (with three zero angles)

points of such curves are always infinite in length, however close together these points are located! But for a century the menagerie of functions has collected so many equivocal and fantastic inmates that we have ended up by becoming acclimated to the deformities of functions without derivatives, which all in all are rather unobtrusive.

Is there any need to emphasize that we consider the division into classical and romantic beauty only a convenient method for understanding and analyzing beauty in mathematics and by no means an absolute and rigid frame? Many of the examples just given have a complex esthetic nature containing side by side several of the categories discussed above.

The romantic wildness of continuous functions without derivatives could evoke in the mystical Hermite the impression that he was battling demons escaped from some mathematical hell. Observe, however, the case of one of these functions, the celebrated Koch curve or *homunculus*. Every arc of this curve, no matter how short, is similar to the entire curve, whose exquisite arabesque it chisels into infinity with unfailing regularity. What could be more classical?[28]

The beauty in mathematical methods will lead us to a still more legitimate distinction between classicism and romanticism, since it concerns the style of human endeavors. This distinction reduces essentially to the antithesis between a desire for equilibrium and a yearning for lack of balance.

Before coming to the actual mathematical methods, it is well to recall that every mathematician possesses a style that can be enjoyed for its esthetic value independently of the scientific results that it permits him to obtain. Whence the well-known parallels between Euler's tactical sinuosities and Lagrange's linear strategy, between Riemann's original lightning flashes and Weierstrass's systematic method of construction, between Kronecker's positivism and Cantor's apocalypse, etc., parallels which for the most part reduce to the inevitable contrast between classicism and romanticism.

CLASSICAL BEAUTY IN MATHEMATICAL METHODS

It seems to us that a method earns the epithet of classic when it permits the attainment of powerful effects by moderate means.

A proof by recurrence is one such method. What wonderful power this procedure possesses! In one leap it can move to the end of a chain

of conclusions composed of an infinite number of links, with the same
ease and the same infallibility as would enter into deriving the conclusion in a trite three-part syllogism.

Certain notations and algorithms have a rare felicity. We must
acknowledge how much we owe to positional notation and the use it
makes of zero. Without it arithmetic would doubtless never have
emerged from its Greek cocoon. And who knows what level civilization
would have been able to reach had it been deprived of this lubricating
fluid? Does not its beneficent influence make itself felt in much of the
mechanism, not only of mathematical technique, but also of those
techniques upon which the power of the great modern states is based?[29]

Vast areas of the theory of functions of complex variables, that
formidable continent discovered by Cauchy, ran the risk of being
passed over almost unnoticed because the appropriate instrument for
exploring them in all their convolutions was lacking. By grafting upon
analysis tissues taken from topology, Riemann surfaces[30] permitted us
to unravel difficulties which would otherwise have remained inextricable, and caused order and clarity to reign there.[31]

Equally classic are the methods which cast a new light on previously known facts, bringing together and unifying discoveries
formerly considered disparate.

On the face of it, the circle and the ellipse have no similarity apart
from the fact that they are both simple closed curves, nor the hyperbola
and the parabola apart from the fact that they are both simple open
curves; and the first two do not seem to have much more in common
with the latter two.

Now, a single definition introducing the notion of focus, in which
only one word is changed (the word addition is changed to subtraction)
permits the close linking of ellipse and hyperbola. A little reflection
will now show that the circle is a special case of the ellipse.

One can go even further and define the ellipse, hyperbola and
parabola together, as the locus of points, the ratio of whose distances
to a fixed point and a fixed straight line is constant and less than,
greater than, or equal to 1, respectively.

When the same protagonists are rediscovered in their different
roles as intersections of a plane or cone, their unification will receive a
new boost. The circle, ellipse, hyperbola and parabola are all now
called *conics*, a name properly given to them to call attention to their
family ties. Projective geometry, of which the plane and cone mentioned above are only a sort of materialization, illuminates still more

profoundly the relationships between the properties of these various figures.

A complete merger was finally established, when thanks to analytic geometry, it could be shown that the equations of the above four curves are all algebraic expressions of the second degree, and reciprocally, that nothing but conics are obtainable from the general and complete second degree equation.

It is hard to imagine that there could be any connection between two curves as different from each other as the Cornu spiral (or cloth-oid) and Fresnel's integrals. However, they both can be obtained by projecting a sort of conical helix, whose equation is rather simple, on three planes each of which is perpendicular to the other two. If this helix is illuminated by suitably directed, parallel light sources, the shadows it projects on planes parallel to its axis will be Fresnel's integrals, and the shadow it projects on the plane perpendicular to its axis will be a Cornu spiral. As a Platonist I could not have wished a better illustration of the allegory of the cave. It goes without saying that this does not provide an argument in defense of the Platonic theses, for our helix has no more and no less reality or intelligibility than the two plane curves whose shadows it forms[32] (p. 140).

Certain mathematical disciplines—most especially the theory of groups, abstract algebra and general analysis—are unrivaled in their aptitude for centralizing; precisely this constitutes their best raison d'être.

We had to wait for the theory of groups to reveal the relationship which unites mathematical phenomena drawn from completely different chapters such as algebra, geometry and analysis. The following example, cited by Emile Borel, has since become classic:

> When Klein makes us see that the algebraic theory of equations of the fifth degree is notably simplified by a prior study of the proper-ties of the regular icosahedron and that this comparison also permits a fruitful study of certain differential equations of the second order, we are lost in admiration at how this overall view illuminates the scattered facts.[33]

We must not think that mathematics progresses only via the royal road of classicism. The directions of research, the scientific ideal and perhaps even the affective climate of certain schools of modern mathematics are without a doubt based on this lucid, passionate and at times somewhat narrow quest for unity. But the dialectics—and also

An Example of a Progression towards Unity: The Conics

. . . these are not spring flowers, at the mercy of the changing seasons, but rather never-fading amaranths, gathered from the most beautiful flower-beds of geometry.

BLAISE PASCAL: letter to M. DE SLUSE

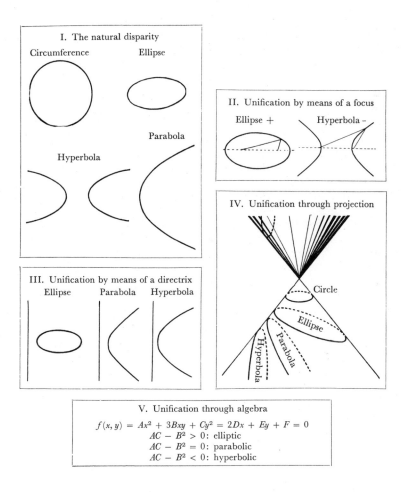

I. The natural disparity

Circumference   Ellipse

Parabola

Hyperbola

II. Unification by means of a focus

Ellipse +   Hyperbola –

III. Unification by means of a directrix

Ellipse   Parabola   Hyperbola

IV. Unification through projection

Circle

Ellipse

Parabola

Hyperbola

V. Unification through algebra

$$f(x, y) = Ax^2 + 3Bxy + Cy^2 = 2Dx + Ey + F = 0$$
$$AC - B^2 > 0: \text{elliptic}$$
$$AC - B^2 = 0: \text{parabolic}$$
$$AC - B^2 < 0: \text{hyperbolic}$$

A certain position of the bolt which positively closes the lock.

PAUL VALÉRY

To Illustrate Plato

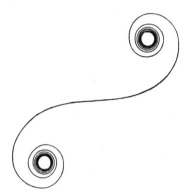

Clothoïde, or Cornu spiral   Fresnel's integrals

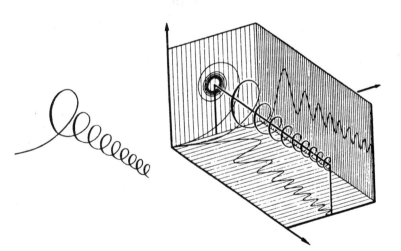

Relief of Fresnel's equation   The objects and their shadows

. . . if they could converse with one another, do you not think that they would suppose that they were naming the real objects themselves when naming the shadows that they were actually seeing?

               Plato

the drama—of mathematical progress consist essentially in an ever-renewed antagonism between this desire for unity and the rebellions which erupt at every new attempt to get to the heart of material reality.

ROMANTIC BEAUTY OF MATHEMATICAL METHODS

Certain types of proofs could be considered romantic because of their indirect character. This would be the case for a *reductio ad absurdum* proof. While just as convincing as other proofs, these actually do not shed any light on the structure of the propositions they establish. A quite characteristic state of dissatisfaction results from this.

There are elementary proofs such as the proof of the irrationality of $\sqrt{2}$ which please as much, if not more, by the nature of the procedures employed as by the facts which result. For example, the proof that the elevation of any real number whatsoever to the zero power always gives 1 merely takes advantage of an ingenious trick. And are not the tricks delightful that Roberval and Pascal employ to calculate the area of a cycloid (by means of wheels which somehow lose their spokes or their chords while turning)? These artifices, which Sergescu reminded us of at a conference in 1947, are proscribed in today's teaching because they revive the excommunicated method of *indivisibles*.

From the pedagogical point of view romanticism (in methods of proof) is very often synonymous with difficulty. The discouragement to which so many pupils yield must be imputed chiefly to the employment of casuistic ruses; while these do honor to the intuitive perspicacity of their authors, they nonetheless strew the labyrinths of comprehension with obstacles.

Elementary algebra—which already normally rebuffs the beginner with its abstractness—could often be expounded with more clarity.[34] It is not always as easy to remain straightforward in elementary geometry. The truths there are skittish; they do not let themselves be approached easily and fall victim only to ambushes whose ingenious inventiveness is far from obvious. Hegel understood this very well, and he stated his criticism—one may not generalize from a single occurrence—in clear language in the two following passages:

> [Construction, in geometry] is thoroughly ordered and one must blindly obey the instructions and draw just the lines in question,

although one could draw an infinite number. . . . After it is done, one also recognizes that these lines were adapted to the end sought, but the adaption to this purpose is only superficial. (*Phenomenologie.*)

Mathematical representation is a tortured representation. (*Naturphilosophie.*)

The proof of the Pythagorean theorem currently used is certainly very clever, with its method of drawing straight lines joining the vertices of the fundamental triangle to the vertices of the squares constructed on the sides, and comparing certain of the new triangles thus obtained. However, the students who cannot assimilate so Byzantine a maneuver have a very good excuse.[35] So also do those students who are confused by the employment of that peculiar technique of calculating the volume of a triangular pyramid by adding to it a quadrangular pyramid whose volume is double that of the initial pyramid, the combined figures forming a readily measurable prism.[36]

There is only too much to choose from in higher mathematics. A few examples will suffice to make apparent the extraordinary divinatory gifts of the great mathematicians.

There is a certain collection of points and straight lines called the *figure of Veronese*, which can be formed from Pascal's *mystical hexagram*, i.e., any hexagon inscribable in a conic. This figure is perturbing enough by itself. It calls for a tropical profusion of elements, one branching off from another, that we could have chosen to describe in the first part of this study. Kirkmans's 60 points, Steiner's 20, Salmon's 15, Pascal's 60 lines, the 20 of Cayley-Salmon, the 15 of Steiner-Plücker, and others still, cross in disorder in this singular jungle. How would one imagine being able to untangle such a wilderness by resorting to a surface of the third degree, and then finding the properties of this latter in a section of hexastigmal space? We call hexastigmal the simplest definable figure having any six points and located in a *four-dimensional space*!

The theory of numbers and *analysis situs*, these high points of arithmetic and geometry, are justly considered two of the altars at which the most secret masses of the cult of mathematics are celebrated. The rites hardly change save for unexpected inventions that are sometimes of spellbinding virtuosity. For example, does it not seem inconceivable that we have to employ the two faces of a single plane to prove the simple proposition that two Jordan curves[37] located on a single surface necessarily intersect in an even number of points?[38]

How can we convey the level of genius that was necessary for certain applications of analysis to the theory of numbers? One is overcome with admiration upon reading Riemann's famous paper "On the Number of Prime Numbers Smaller than a Given Number (1859)." This introduces that ideal catalyst in the laboratory of prime numbers, the fantastic zeta function (p. 145).

Seeing a certain method succeed regularly every time it is employed makes it more disconcerting to discover its inefficacy in a new instance, and increases all the more the enthusiasm with which we greet an entirely original and successful procedure. This is what occurs, for example, when the theory of composite functions is employed to calculate certain derivatives, that of $y = x^x$ among others.

Extraordinary acts of daring accompanied the birth of elliptic functions. Why did elliptic integrals resist the assaults of the best mathematicians at the end of the 18th and the beginning of the 19th centuries? Because analysts of Legendre's caliber did not think of anything besides timeworn tactics to storm this citadel. How could one doubt that even more darkness would be cast on these questions by investigating the *inverse* functions of these recalcitrant integrals, and particularly by shifting them from the real to the *complex* field? Far from aggravating the difficulties of the problem, this audacious double twist permitted the magical trumpet blast that sent the walls of this mathematical Jericho tumbling down. Abel and Jacobi victoriously entered the city and set about delivering the keys to a treasure whose value would be shown by the energetic development of the theory of functions in the 19th century.[39] Mittag-Leffler writes:

> The best works of Abel are true lyric poems of sublime beauty, whose perfection of form allows the profundity of his thought to show through, while at the same time filling the imagination with dream visions of a remote world of ideas, raised farther above life's commonplaces and emanating more directly from the very soul than any poet, in the ordinary sense of the word, could produce.

THE BEAUTY OF MATHEMATICS
AND OF ITS DEVELOPMENT

The flaming iridescence of a rose window in a Gothic cathedral should not hide from us the grand architectural unity of the whole. In the same way there is a geometry of mathematics as a whole; it is in its totality that it is beautiful, with a splendor surpassing the most ineffable visions.

No other area of human thought or intelligence operates with such light-headed intrepidity. E. T. Bell could write, "The essence of mathematics is its eternal youth." And Gonseth does not hesitate to declare, "It is not a paradox; the spirit of adventure, a sort of heroism, animates the mathematician far more than his formulas."

An epic breath unceasingly swells and expands the notions of number, space and function that have been outlined in this study. This fluttering of wings in the sky of abstraction always upsets the best minds. Aside from many declarations of a sometimes mystic cast,[40] the scars this has left are to be seen in the very names given the successive generalizations of whole numbers: irrational, complex, ideal, transcendent, etc. Do not these designations reveal a rather unscientific state of mind seeking to eternalize esthetic emotions rather than the elements of a clear definition?

We know how, by repeated generalizations, geometers have increased the value of their initial capital, the usual space of three dimensions. With an eye on algebraic models they first framed the concept of four-dimensional space. Then by repeating the same operation over and over, they produced spaces of ever-increasing dimension, ending up by passing beyond the geometry of n dimensions (n being any number whatsoever, variable or unknown) to the geometry of spaces with an infinite number of dimensions. Why then be limited to a denumerable infinity? It took but a little time to cross this Rubicon and fashion spaces with dimensions greater than infinity.

If a romantic excess seems to characterize the style of these efforts to outbid their precursors, we discern a classic order in the famous "*Erlangen program*" proposed by Klein in 1872.[41] This classification of particular geometries, which constitute as it were the important ministries of the empire of geometry, associates with each geometry a group of transformations that characterize it. As we go from a group of transformations to a more general group, which includes the former as a subgroup, we correspondingly go from one geometry to a more general geometry. Thus, by successive distillations we extract from metric geometry—which preserves translations, rotations and symmetries—Euclidean geometry, which preserves similarity properties; then, projective geometry, which preserves properties under projections and sections; then algebraic geometry, which preserves

The terrifying realms of worlds in formation.

JEAN PAUL RICHTER

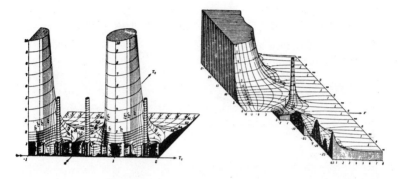

Relief of the elliptic modular function

Relief of Riemann's zeta function

Yves Tanguy—"Suspended Time"

POLYHEDRA

Jacopo de Barbari—Portrait of Fra Luca Pacioli and Duke
Guidobaldo
(1445?–1511?) Museum of Naples.

properties under birational transformations; and finally topology, which preserves properties only under homeomorphisms.

These are solemn moments of prestigious and significant beauty, when disciplines up to then distinct establish contact with each other and, making various assorted matches, are wed, each one preserving its own individuality even while being absorbed into a higher unity.

A new breath was infused into the whole of ancient thought when arithmetic met geometry under the sign of Pythagoras. Let us listen to Edgar Quinet celebrate the marriage of algebra and geometry under the sign of Descartes:

> If I was smitten by algebra, I was dazzled by the application of algebra to geometry. . . . The idea, the possibility of expressing a line, a curve, in algebraic terms, by an equation, seemed to me as beautiful as the *Iliad*. When I saw this equation function and solve itself, so to speak, in my hands, and burst into an infinity of truths, all equally indisputable, equally eternal, equally resplendent, I believed I had in my possession the talisman which would open the door of every mystery.

The marriage of the theory of functions and the theory of surfaces under the sign of Monge is of the same character. And calling to mind the introduction of the theory of groups of substitutions into the domain of algebraic equations under the sign of Evariste Galois, the introduction of the theory of groups of transformations into geometry under the sign of Felix Klein, and the merger of these two theories with the theory of abstract groups under the sign of Sophus Lie, are not these sufficient to make us believe that we are hearing the three fateful knocks which announce the raising of the curtain upon the drama of 20th-century mathematics?

This is how beauty evidences itself in mathematics, as in the other sciences, in art, in life and in nature.

Sometimes the emotions it stirs up are comparable to those of pure music, great painting or poetry, but usually they are different in character, and can scarcely be comprehended by one who has not felt their glow within him. To be sure, the beauty of mathematics guarantees neither its truth nor its utility. But to some it brings the gift of being able to live matchless hours, to others the certainty that mathematics will continue to be cultivated for the greatest good of all and for the greatest glory of the human adventure, by men who expect no material profit for themselves.

NOTES

[1] It is worthwhile looking for an explanation of this astonishing precocity on the part of so many great mathematicians. One scarcely finds its equivalent except in music or chess. Think, indeed, of the great German geometer Steiner, who had discovered theorems about the properties of the sphere at thirteen; think of Pascal, rediscovering Euclid's *Elements* at twelve *with the aid of balls and rods*; finally, think of Gauss the most prodigious of them all, of whom it is said that he solved numerical equations at the age of three and a half. How can one conceive of the attraction that mathematical truths could exercise upon such very young children if not by the incitement of their imagination and of their desire to play joined to the exercise of their mind?

In a sense relating to this, we can recall the anecdote which Pascal's niece, Marguerite Périer, tells in the following words:

> . . . he got a very painful toothache. One evening the Duke of Roannez left him suffering in agonizing pain; he took to his bed, and inasmuch as the pain only increased he decided that for relief he would apply himself to something that could make him forget the aching. To that end he bethought himself of the proposition concerning roulette, stated previously by Father Mersenne, but which nobody had been able to prove, and upon which he had never spent any time. So productively did he think about it that he found the solution and all the proofs. This serious industry diverted his attention from his toothache and when he stopped thinking about the proposition he felt himself cured.
>
> When Monsieur de Roannez came to see him in the morning and found his toothache gone, he asked what had cured him. He responded that it was the roulette, which he had sought and found.

"Roulette" was Pascal's name for the cycloid we are going to study in note 4.

[2] This magic square of order 4 satisfies a supplementary condition as well: the two numbers 15 and 14, situated side by side in the middle of the bottom row, recall the year when this masterpiece was engraved.

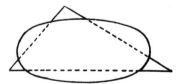

[3] But we will have to address ourselves to the complete geometry of the triangle. Let us refer once more to Carnot's theorem relating to the inter-

section of a triangle and any ellipse which cuts it in six or more points. These six points divide each of the three sides into three segments, and if the three segments located in the middle of each side are disregarded, the six others distribute themselves into two groups of three, the products of which are equal.

It can be shown in projective geometry that the 24 anharmonic ratios that exist among four points can be grouped according to six ratios between the number 1 and another suitably chosen number. The initial variety is controlled by a sense of architectural regularity.

[4] It is the curve—formed of successive arcs—described by a point on a circumference which rolls on a straight line without sliding; the trajectory described by the air-valve on a bicycle tire is a cycloid. The chord which subtends the cycloid has a length that is obviously equal to the circum-

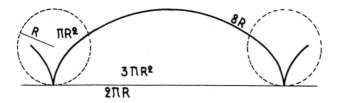

ference of the generating circle, $2\pi r$ if one designates the radius of the latter by r. In this case the length of the cycloid itself is $8r$ and the surface included between the cycloid and its chord is three times that of the generating circle, or $3\pi r^2$.

The cycloid was hinted at by the Cardinal of Cusa and Charles Bovelles in the 16th century. It was investigated by Galileo, Roberval, Pascal and others in the 17th century.

[5] Imagine a clock whose minute hand turns, as usual, with constant

angular velocity. Starting at the center, an insect advances along this second hand at a velocity which increases in geometric progression. If one imagines that the insect's tail has been dipped in ink, the tracing it leaves on the clock's face will be a logarithmic, or equiangular, spiral.

[6] Spherical trigonometry abounds in similar formulas. A heteroclite assemblage is repeated several times so as to form a structure in which slides, turns and permutations of parts of the original element end by making this complexity yield to the simplicity of a higher order.

[7] This insobriety was certainly not unconnected with the fact that Gauss was dissatisfied with only one kind of proof of this proposition stated by Euler and hinted at by Legendre. Several times in his life he returned to "his theorem," and he left seven different proofs of it. There are now more than 50.

[8] Here is the triangle. It needs only a trifling examination to discover the law of its formation.

$$
\begin{array}{l}
1 \quad 1 \\
1 \quad 2 \quad 1\cdot \\
1 \quad 3 \quad 3 \quad 1 \\
1 \quad 4 \quad 6 \quad 4 \quad 1 \\
1 \quad 5 \quad 10 \quad 10 \quad 5 \quad 1 \\
1 \quad 6 \quad 15 \quad 20 \quad 15 \quad 6 \quad 1 \\
1 \quad 7 \quad 21 \quad 35 \quad 35 \quad 21 \quad 7 \; 1 \\
1 \quad \dots\dots\dots\dots\dots\dots\dots\dots \\
1 \quad C_m^1 \, C_m^2 \, C_m^3 \, C_m^4 \, C_m^5 \, C_m^6 \dots\dots C_m^m
\end{array}
$$

[9] "Newton's binomial" is the name given to the following formula, which gives the development of the sum of two terms, x and a, raised to any given power:

$$(x + a)^m = x^m + C_m^1 \cdot a \cdot x^{m-1} + C_m^2 \cdot a^2 \cdot x^{m-2}$$
$$+ \cdots + C_m^p \cdot a^p \cdot x^{m-p} + \cdots + C_m^{m-1} \cdot a^{m-1} \cdot x + C_m^m a^m$$

The coefficients C_m^p are those of the preceding note.

[10] In *combinations* of m objects taken p at a time, the order in which the objects are arranged is disregarded; in *permutations* it is taken into account. In both cases one may or may not consider repetitions. Combinations without repetition of m objects taken p at a time are the numbers C_m^p in Pascal's arithmetic triangle (see note 8).

[11] See particularly the series expansions in Dubreil's and Brunet's articles. Let us again state Fagnano's formula:

$$\pi = 2\sqrt{-1} \log \frac{1 - \sqrt{-1}}{1 + \sqrt{-1}}$$

and this form of Stirling's formula:

$$\sqrt{2\pi} = \frac{n!e^n}{n^n\sqrt{n}} \cdot \frac{12n}{12n - \theta}$$

provided that n approaches $+\infty$ and $0 < \theta < +1$ ($n! = 1 \times 2 \times 3 \times 4 \times \cdots \times n$).

[12] See particularly the series expansions in Dubreil's article.

[13] Taylor's formula gives the series expansion of $f(x + h)$. If we let $x = 0$ and $h = x$ we get Maclaurin's formula:

$$f(x) = f(0) + \frac{x}{1!} f'(0) + \frac{x^2}{2!} f''(0) + \frac{x^3}{3!} f'''(0) + \cdots + \frac{x^n}{n!}(0) + R_n$$

We shall not give explicitly this last term R_n. It tends toward 0 when n tends toward infinity.

[14] Here are a handful of examples:

The strophoid is a transformation by inversion of an equilateral hyperbola.

The orthogonal projection of a Maclaurin trisectrix is a folium of Descartes.

The locus of the points of inflection of Sluse's cubics having the same asymptote and the same singular point is a cissoid of Diocles.

The axis of a parabola inscribed in any triangle envelops a hypocycloid of Cremona.

Archimedes's spiral is the polar of the involute of the circumference (with respect to the center of this circumference).

The conchoid of Cappa is a Mascheroni curve.

An ellipse A, a hyperbola of Appollonius B whose center describes the ellipse, and the parabola of Chasles C, the polar reciprocal of B with respect to A, form a harmonic system; the envelope of B is the derived Kreuz curve derivative of the ellipse; the envelope of C is the evolute of the ellipse whose cusps are the vertices of the ellipse A.

We could extend this picture of geometric correspondences indefinitely. And that is why we end up feeling surfeited. However, even when the edge has been taken off our capacity for surprise at such highly advanced transformations, we still remain susceptible to the revelation of the Lie transformation (straight lines into spheres) mentioned in Elie Cartan's article, Part II, Book One, pp. 262–267.

[15] Differentiation and integration are the two fundamental operations under whose vaults spreads the vast symphony of infinitesimal analysis. See note 8 of Brunet's article, p. 255.

[16] These are obtained by making a single twist in a paper ring. If one

traces a line along what might be called the axis of the ribbon, one soon encounters "the other side" of the strip, i.e., one confirms that this ribbon has only one edge (or side), and in addition does not possess a top and a bottom, but rather a single face. But this is not the only entertainment we can derive from this geometric sphinx. If we make an incision

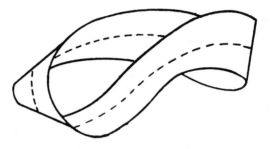

along the line previously mentioned, we are surprised to see that only one ring instead of two is obtained. If we perform the same operation again on the ring thus obtained, there is another show-stopper; we have made two rings appear this time!

On the same order of ideas, let us also mention "*Klein's bottle,*" a closed surface which possesses neither interior nor exterior.

[17] Although a straight line is the path of *least distance* between two points, it is not a brachistochrone, i.e., the path of *least time* between two points located at different levels in a space governed by gravity. This honor belongs to the cycloid. Under the same conditions every portion of the path thus defined is covered by a moving weight in the *same time* (tautochronism) as the total path (see page 295 of Vol. I).

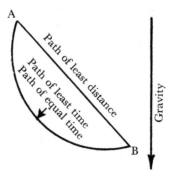

¹⁸ Through a point exterior to a line, one can draw one and only one line parallel to the given line.

¹⁹ Through a point exterior to a straight line one can draw an infinity of straight lines parallel to the given line (see the article by Sainte-Lagüe, Vol. I, p. 130, and by Thiry, Vol. I, p. 145).

²⁰ Through a point exterior to a straight line, one cannot draw any parallel to the given line (see the articles by Sainte-Lagüe, p. 130, and by Thiry, p. 145).

²¹ On a horizontal plane let us draw parallel lines equally distant from each other. Let us randomly throw, a great number of times, a needle whose length is equal to the distance between two consecutive parallels.

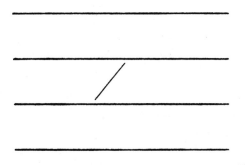

Let us count the number of times that the needle falls between two consecutive rays without touching them and the number of times that it falls across one of the rays. The greater the number of trials, the closer is the quotient of these two numbers to π.

²² The probability that any two given integers are prime with respect to each other is equal to $6/\pi^2$ (Cesaro and Sylvester, 1883). That is, if one chooses integers at random and pairs them, two by two, the quotient of the number of pairs of integers that are not prime with respect to each other divided by the number of pairs of integers that are prime with respect to each other will be closer to $6/\pi^2$, the greater the number of pairs tested.

²³ Note with special attention the articles by Denjoy, Frechet, Eyraud and Buhl in this book. For many mathematicians the theory of groups constitutes the logical point of departure for all of mathematics.

²⁴ *Matrices* is the name given to rectangular tables of numbers whose raison d'être appears in various chapters of mathematics (tables of coefficients of systems of algebraic forms, of systems of linear substitutions, etc.). In turn they find important applications in the most modern

branches of physics. We can define addition, subtraction, multiplication, division, differentiation, integration, etc. of matrices, just as we do for numbers. The product of two matrices is a new matrix which is obtained by multiplying the rows (horizontal) of the first by the columns (vertical) of the second! At first glance doesn't this appear over-elaborate, like a Baroque decoration?

[25] To understand the pseudosphere, we must first describe the tractrix. The tractrix, or "curve with equal tangents," is a plane curve whose definition we shall not give in terms of a differential equation, Cartesian equation or characteristic geometric property. We shall content ourselves with indicating how one can obtain it mechanically. Let us place on a

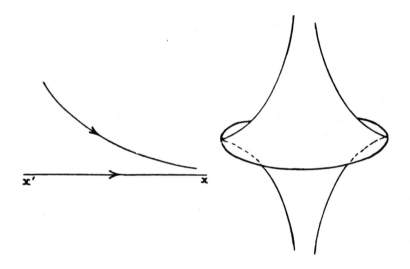

horizontal table a small object that is not on the straight line $x'x$ and let us attach to it a flexible thread of fixed length. The other end of this thread shall be located on the straight line $x'x$. The thread is drawn along the straight line in the direction $x'x$. The object pulled by the thread describes a tractrix as it approaches the straight line.

The pseudosphere is generated by the revolution of a tractrix around the straight line $x'x$ which served to generate it. It is a surface with negative curvature. As Beltrami has shown, it plays a large role in the interpretation of Lobachevsky's non-Euclidean geometry. Although the tractrix extends indefinitely (for it never meets its directrix $x'x$, which is asymptotic to it), the surface and volume of a pseudosphere are finite. The volume of a pseudosphere is half the volume of the sphere having the same "equator." Its surface is equal to that of the same sphere. We know that the quadrature of the circle is an impossible problem. Transposed from the plane to the pseudosphere the same problem becomes possible.

[26] An analogous shock is produced when we compare the notion of radius and diameter in plane geometry. To say that the circle is a figure all of whose radii and all of whose diameters are equal rightly seems a

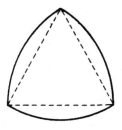

redundancy. We had to wait until 1875 to realize that while the circle remains the only plane figure all of whose radii are equal, it does not have a monopoly on the equality of all its diameters. Other figures, the figures of Reuleaux, can claim the same honor. They are readily constructed by starting with regular polygons possessing an odd number of sides, and "replacing" the sides by arcs of circles whose centers are the opposite vertices.

[27] For example, in a plane—a two-dimensional space—our figure is a circle, and its measurement, which is an area, is πr^2. In ordinary space—of three dimensions—the figure is a sphere, and its measurement, which is that of a volume, is $\frac{4}{3}\pi r^3$. In "extension"—a name sometimes given to four-dimensional space—the figure is called a "hypersphere," and its measure, which is that of a "hypervolume," is $\frac{1}{2}\pi^2 r^4$. We would then have, for five dimensions, $\frac{8}{15}\pi^2 r^5$; for six dimensions, $\frac{1}{6}\pi^3 r^6$; for seven dimensions, $\frac{16}{105}\pi^3 r^7$; for eight dimensions, $\frac{1}{24}\pi^4 r^8$; for nine dimensions, $\frac{32}{945}\pi^4 r^9$, etc.

[28] On the other hand a single chapter of mathematics can present a veritable mosaic of styles. We find this in the study of regular polygons and polyhedra, and also in their generalization, the study of regular figures of n dimensions.

A formula due to Euler informs us that the sum of the number of vertices and the number of faces of any polyhedron is equal to the sum of the number of edges plus 2. How well designed these edifices are! On the other hand, let us examine and enumerate the "regular figures" which exist in spaces of 2, 3, 4, 5, ..., n dimensions. As we know, an *infinite* number of regular polygons exist in the plane. And, in space, there are only *five* regular polyhedra: regular tetrahedron, hexahedron (or cube), octahedron, dodecahedron and icosahedron. If we penetrate into four-dimensional space, we discover *six* regular polyhedroids: regular octahedroid, pentahedroid, hexadecahedroid, hexacosihedroid, icosatetrahedroid, and hecatonicosahedroid. Let us pursue our inquiry. From here on we will find only three regular figures in spaces of more than four dimensions, no matter what number of dimensions the space considered has. Why the irregularities of two, three and four-dimensional spaces before we come upon the subsequent arid monotony? However, this anomaly is only apparent; we find the clue to the secret in the light of the correlations which can be discovered among the elements of the various figures.

And how is one to define the sort of beauty which can be claimed by the theory of regular polyhedra when it insinuates itself into certain sectors of the integral calculus in order to organize them? Buhl in his *Nouvéaux Eléments d'Analyse* writes:

> That the problem of possible and relatively elementary integration comes down to the regular polyhedra is of the highest philosophical importance. What geometer has ever contemplated actually constructing regular polyhedra? It seems that this artistic work is to be left to a jeweler who knows how to cut precious stones. Now, the search for mathematical harmonies, in this instance the harmony accruing from the existence of certain possibilities of integration, partakes of the same artistic quality.

[29] Logarithms. These operators possess a power which their simplicity increases tenfold, and which makes their employment so valuable in the domains of numerical and analytical calculus. If the slide rule, based as we know on logarithms, did not exist, modern industry would probably need twice as many engineers.

[30] See the article by Valiron, Part I, Book Two, p. 167.

[31] Another example: The functions of complex variables become infinite at a certain number of points of a plane; this seems to rule out any summation of their value on this plane. A method has been devised whereby these troublesome spots are encircled with "laces" which can be pulled in as tightly as one wishes. The combination of the infinitely small circumference of each of these circular prisons with the infinitely large value taken by the function that it encloses allows us to rid the desired summation of the supposed difficulties.

[32] This may be compared with the example of the shadow of the five divergent parabolas given by Brunet, Vol. I, p. 260.

[33] In a somewhat related class of ideas, general analysis, which owes so much to the work of Fréchet, has permitted us to start from one central

focal point and deduce results which then spread like bountiful manna over the most diverse and remote regions of mathematics.

It is well known that general analysis, thanks to the minimal hypothesizing which it strives to get by with, makes possible the simultaneous study of seemingly entirely different mathematical subjects. Through this it makes a contribution to the unification of the science and to the apprehension of profound analogies between subjects that at first glance are unlike. Thus the polynomials of elementary algebra, multilinear forms possessing a finite or infinite number of variables, homeomorphism isomorphisms of one Abelian group on another, the first members of numerous differential and integral equations, integrals, etc., are particular realizations of abstract polynomials. (Van der Lijn.)

[34] The path generally taken to solve a second-degree equation becomes clear, and in some way classic, only when the student gets over the impression of having been shown a conjuror's trick and perceives the main idea. It consists in the comparison of the second-degree equation with the expansion of the square of a sum and in the attempts made to adjust and perfect this resemblance.

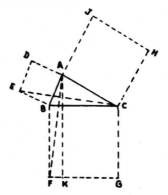

[35] Up to a certain age, one could be satisfied with a less rigorous proof, conceived in India, which was occasionally used. It is quite sufficiently convincing especially if it is worked out on graph paper.

On the other hand we admire the pure and elegant classicism of the proof of a wholly comparable theorem relating to similar right triangles constructed respectively on the hypotenuse and on the two other sides of any right triangle. If one erects the altitude on the hypotenuse (remembering that this altitude divides the initial triangle into two similar triangles) it is immediately evident that the large triangle is the sum of the other two.

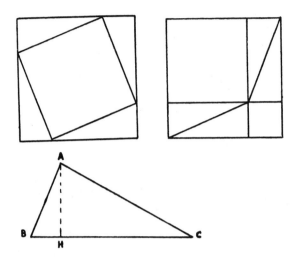

³⁶ In this study we are not going to examine the question as to whether it is desirable to let certain carefully selected difficulties remain in elementary instruction for the purpose of spurring on and strengthening the student. There is no doubt that mathematics, which is above all a *science* and a *technique*, is also not only—as we hope we have shown—an *art*, but equally a *mental gymnastic.*

³⁷ We shall not give a rigorous definition of a Jordan curve. We can get a sufficiently good idea of it by drawing as twisted a closed curve as we wish on a plane, providing it does not intersect itself, and then crumpling this plane or even stretching it like a rubber sheet, being careful not to tear it. See Lentin's article, p. 205.

³⁸ Counting two points of tangency.

³⁹ See on this subject Valiron's article, p.162 , and Godeaux's, p. 287.

⁴⁰ "The Divine Spirit has found a magnificent creation in this marvel of analysis, in this framework of the ideal world, in this kind of amphibian, neither being nor non-being, which we call the imaginary root of negative unity" (Leibniz).

"There is something hidden there which is incomprehensible to us" (Huygens, apropos of complex numbers).

⁴¹ One could almost measure the importance of this "Erlangen Program" by the number of times it has been alluded to in this book.

41

THE ESTHETICS OF SCIENCE AND MODERN THEORIES

by Adolphe Buhl

HONORARY PROFESSOR OF THE SCIENCE FACULTY OF TOULOUSE

IT is truly an intriguing thing for a mathematician about to terminate his university career to consider the extraordinary transformations undergone by mathematical thought during this career. It is scarcely half a century since we became engaged in investigating the foundations and fixed principles upon which geometry and mechanics were to rest. Not without warrant, mathematicians wished to establish these sciences on the basis of half a dozen postulates which, once chosen, would be fixed in character, further progress to consist only in developments concerned with the superstructure and not with the foundations of the scientific edifice. Some even declared it desirable, from the point of view of teaching, to erect foundations that, once obtained, would never require any reconsideration. I do not exaggerate. I have heard such wishes expressed at conferences and at a goodly number of meetings of scientific societies.

Not everybody was convinced. The opposition, of whom I was one, saw a kind of mysticism in this. It is certainly true that the human mind often evinces a great reluctance to change its principles, to learn something different from what it had already learned. Religions have pursued dogmas declared to be immutable and yet not all believers took them to be so. Furthermore I have heard preachers in their pulpit agree that present times call for new study of fundamental religious concepts.

Whatever it may be, science has not been dogmatic. Whether wished for or not, new scientific demands have led it to reconsider its principles, to alter them, and sometimes to assist in their exit without this in the least destroying the possibility of research and discovery.

Thus rational classical mechanics may possibly seem to rest on some postulates, but it is absolutely inadequate in the corpuscular world where the wave mechanics created by Louis de Broglie acts a part. We have had to replace Jacobi's first-order partial differential equation, which plays a satisfactory and admirable role in celestial mechanics, with Schrödinger's second-order equation; this permits us to establish connections between waves and corpuscles. Having come this far, it is easy to see that there are equations of any order n which permit the mastery of such liaisons. To be sure, in the case of the integer n we are in the presence of more and more perplexing complications, and we must counsel the neophyte to stick with the case of $n = 2$ for a long time before getting ready to go on. But there is superabundant proof that we will never find a unique, fundamental mechanism to interpret corpuscular phenomena. On the contrary, we must wait for what reflection and work always bring to a rethinking of a subject, especially when new phenomena come to our attention. The very word *mechanism* has something imperfect about it that must arouse distrust. A mechanism is constructed with ordinary matter, i.e., with myriads of corpuscles. How slight the probability that average and gross properties— the perceived properties—will still be those of the isolated corpuscles.

Non-Euclidean geometries evoke analogous concepts. There are an indefinite number of these geometries, for they arise from the properties of *groups*—a vastly varied and infinite collection of them. One must resign oneself to never knowing more than very little about them, but the *set* of these groups does exist, and rules out all definitive knowledge of the geometry of the universe. New observations may lead precisely to those geometric domains that were originally overlooked.

To be more specific, it is easy to show that if there is a corpuscular geometry, then it cannot be Euclidean.

First of all, what is Euclid's set of postulates? We could present manifold statements concerning this, the most remarkable perhaps being the one affirming the existence of *similar* figures, i.e., figures possessing the same shape but having different dimensions. This postulate is a great aid to science. For example, it permits one to represent a minuscule insect, say a microbe, on a page of a biology book—*greatly enlarged*. But has anyone anywhere drawn an electron *greatly enlarged* in order to show its inner structure? The answer is no.

One felt the impossibility of the task, although one did not explain it very well to himself. However, there have been attempts in the manner of Bohr to depict atomic structures with a central nucleus and electronic trajectories. But Bohr himself has advised rejecting such images, which can only give rise to false ideas. In this instance the very notion of an image is unacceptable; every image supposes taking an extremely large number of corpuscles from the ink, the pencil and the paper. How strange this corpuscle which would then have to be represented by so many of its fellows!

In short we have the same difficulties with *image* as we have with *mechanism*. The difficulties seem insurmountable, but again we must not give ourselves new wrong ideas as to any insurmountability. When questions appear not to have an answer, we must ask whether these questions have a meaning. Now mankind possesses the peculiar faculty of being able to pose an indefinite number of questions having no precise meaning. How many givens we take from one domain and transport to another without perceiving the absurdity of the transfer!

It required relativity to show us that not everything in ordinary space and time was explicable on the basis of common sense. And yet what more or less dim-witted rebellions there were against the generalizations of a scientist marked by superior intelligence!

But will intelligence itself, supposing that it is not harried, combatted or persecuted, but enjoys full freedom, finally bring us some glorious serenity? Alas, no. After a life devoted to the most elevated speculations on analysis, the illustrious David Hilbert sought to crown his work with a general theory of mathematical reasoning. In the main, his efforts, aided by his best disciples, seem to have led him to the position of seeking to justify reasoning by means of reasoning!

The theory of sets, to which I alluded above, contains at least as many difficulties as wave mechanics and yet it is generally believed that it should be the basis of all mathematical theory.

We must guard here against intentionally playing paradoxical games; the true scientific attitude is just the opposite and consists in attempting to reduce the number of paradoxes. But to the common mortal, sets at first glimpse seem no less pregnant with germs of discord. For example, by stating that the prototype denumerable type of set, the set of integers

$$1, 2, 3, 4, 5, \ldots \qquad \text{(a)}$$

and the set of multiples of 10, say

$$10, 20, 30, 40, 50, \ldots \qquad \text{(b)}$$

have the same *power*, the same *wealth* of elements, I have often pro-voked vehement contradiction.

Let my readers try, even if it means making a parlor trick out of this. I pointed out that writing the two lines one under the other, element for element, showed clearly that to *one* element of the first line, there corresponded *one* element of the second line. The rejoinder of some was that in line (b) I did not write the integers from 1 to 9, from 11 to 19, from 21 to 29 and so forth, integers which did exist in line (a).

Therefore set (b) was less *complete*, less full, less *powerful* than (a). Oh well, such protesters should not be judged disparagingly. I have always felt their sincerity to be genuine. I know of only one way of defending the equipollence of (a) and (b). This is by saying that this equipollence launches a theory that can then be developed indefinitely in an eminently useful and esthetic way, whereas the denial of equi-pollence, though probably in itself defensible, does not lead to any development of value. But this argument might not be convincing, for in order to appreciate its worth one must first have gone rather deeply into the theory of sets.

If we move ahead to the notion of *measure* of a set, a concept due to Emile Borel, we soon come upon strange-looking new objects. Sets possessing the power of the *continuum* can have *zero* measure. Where there is nothing to be measured, possibilities of *structure* can still exist.

On the same order of ideas is the development of *abstract* trans-formations.

Many minds rebel at thinking of transformations wherein we do not know what is being transformed, and yet this has put mathematics in argument with the old philosophic view according to which we are not permitted to know the "underlying foundation of things" but only changes. Again, the "underlying foundation of things" is very probably a meaningless expression. Such a foundation would be a phantom created by an imperfect human understanding. The most reasonable idea to associate with this imperfection is assuredly the notion of *group*, which we have already mentioned; i.e., the notion of trans-formations linked in such a way as to form a whole that is necessarily coherent because it always contains elements of the same nature as those already linked.

All classical physics comes from the concept of measurement.

Measuring lengths, areas, volumes, extensions in n dimensions means writing multiple integrals about which it is sufficient to know some transformation laws in order to arrive at equations for phenomena, such as Maxwell's equations in the electromagnetic theory of light. The completion of these equations gives the gravitational phenomena. And this constitutes Albert Einstein's gravitation theory, which thus also becomes dependent upon the notion of measurable extension.

Now we are accustomed to these ways of constructing a science, but what resistance there was at the outset! What, they asked, you want to construct a world of physical phenomena with nothing but geometric space? Ah, well. The space that seems abstract to you is very physical; it is obviously there, and you believe you *see* it; yet you analyze it only with the intervention of extremely complex cerebral phenomena. Compared to these, the problem of gravity is a small matter.

Wave mechanics, too, has an operational structure, but it is less dependent upon the notion of measurement than is Einstein's gravitation theory. Its corpuscles resemble as little as possible the material points of rational mechanics. Even in general they are not *localizable*; at best we can say that at time t there is a certain probability that they will be in certain spatial cells. But how do we interpret time t itself? Is it astronomical time? How does celestial motion come into the study of the atom? All these questions have been differently viewed by different theoreticians. No, there is no reason, always speaking generally, for the atom to borrow anything from a heavenly body. Would not the contrary be likely, rather? Thus, t can only be a parameter indicating succession, in the most general case without the possibility of *repetition*. For when we say that a phenomenon is *repeatable* we are implying that it is so in astronomical time.

The basic technique of wave mechanics is the *algebraic matrix*, i.e., the symbol of a linear transformation. Since we can apprehend only transformations, let us begin our theoretical constructions with the simplest symbols, which are those that indicate a *linear* relationship among variables (i.e., relationships of the first degree). This is how a *matrix* mechanics arises. Its differential equations are preserved under linear transformations and these transformations are extended (*representation* theory) so as to correspond with the spectra of different substances. Upon this spectral canvas appear the modalities of phenomena more or less comparable to motions.

But here again, we must not misuse the word. The concept of the classical material point is absolutely without value in all this; it follows that the trajectories of a material point no longer have any value. A corpuscle that changes position ought to have a certain locus, but this cannot be an ordinary curve accompanied by tangents and notions of continuity better suited to fashioning an imaginative geometry than a real physics.

The corpuscular trajectory must have a *fine-* or special *micro-structure*, and there is every reason to believe that to a corpuscle of a certain species there correspond equally special trajectories. One cannot assign to these a precise position accompanied by an equally precise velocity, any more than we can have a precise velocity accompanied by an exact localization. This is one form of *Heisenberg's uncertainties*.

The combination of such considerations makes for a complex science whose difficulties grow without end. The 19th century subsisted on simple ideas disturbed, only towards the end, by the discovery of radioactivity. Now we are in a period of complexity which seems to be following the preceding period in the manner predicted by Henri Poincaré: We find complexity in simplicity and expect to find a new simplicity beneath the complexity, and so on. But an Ariadne's thread to help us move with ease in wave mechanics has not yet been found.

Other complications arise from the *quantum* character of the new mechanics.

In elementary mathematics we learn as beginners to admire the generality that algebra possesses by virtue of its indeterminate coefficients; change the values of these coefficients, and the methods of reasoning will remain the same. But as soon as one has entered the realm of differential equations, and even more so, that of partial differential equations, the least change in the coefficients of an equation entirely changes the nature of its solutions. Once more, it is impossible to follow the changes by means of continuity. One tries different ways of following them, in particular *quantification*, but a new source of difficulty then arises, inherent in the physical as well as the mathematical nature of things.

Continuity was definitely a simple, not to say simplistic, invention, and in any event a very deceptive one.

We have yet to question ourselves about the value of all this new science; and we shall do this mainly from a philosophical point of view.[1]

First of all, we believe in *esthetic* value. Combinations of operators, matrix properties, mechanical syntheses on spectral canvases are sufficient to arouse a lively appreciation of the esthetic values.

In general we do not know exactly what can qualify for the adjective *true*; there is only the possible in ever-varying forms. The *beautiful* is no easier to define, but it possesses the power to give such great satisfaction that doubtless there is nothing better that we can ask for.

We now possess a mathematics that one can truly see as a reflection (sometimes multiple in aspect) of the divine thought that created the universe. The concept is Hellenic, but within the last century such men as Maxwell, Einstein, Lie, Poincaré, Cartan and Louis de Broglie have rejuvenated it. And all this has taken place without a claim from any of these geniuses that he has given the final word. On the contrary, they have all warned us against the vain search for an absolute. The multiplicity of scientific systems corresponds to the multiplicity of religions, but whereas the religious individual generally believes in the superiority of *his* religion, the true scientific spirit grants the relativity of the different systems, although certain scientists are too stubbornly attached to their conceptions.

Since I have just made a comparison between science and religion, I must mention that the latter, entirely to its credit, seems to have understood the vanity of such notions as time and space, which science once found fundamental, well in advance of modern scientific developments. In any event, God is not localized in ordinary space; and the extreme paucity of the notion of time has been countered by the concept of an *eternity* which can be defined as the totality of everything not dependent upon time. Beings of light, for whom time and space would not matter, were envisaged long before Einstein imagined a physical example of such a being.

Astronomical or ordinary time often seems to me to have an extremely disturbing imperfection. In conjunction with the biological idea of evolution, it gives rise to many meaningless statements, and far too readily. I believe in biological evolution but on condition that no attempt is made to place it completely in astronomical time. I admit

[1] The reader who would like to study the mathematical aspect of matter can find the essentials in our *Nouveaux éléments d'analyse* (Gauthier-Villars, Paris). This work is written specifically to bring classical analysis into relation with new needs, especially those pertaining to gravitation theory and wave mechanics.

that we are products of evolution, and this seems to situate us in a sensible universe where temporal forms appear as particular objects, just as astronomical space is only a particularization of more general spaces. Next we would like to see a decidedly more general development with reference to these cases of space and astronomical time, which are too specific and too material; conceptions departing considerably from the cases in question would then be possible.

There is a *biological* time. To my knowledge the author who has most skillfully defended this position is Alexis Carrel in his admirable book *Man, the Unknown* [*L'Homme, cet inconnu*].

Historical time does not seem distinct from astronomical time; it is marked off in the same numerical way, by dates.

History, is another very interesting science linked to the concept of *progress*.

Must we believe in progress, particularly in the progress due to science? The present period scarcely favors such a concept. War is more and more barbarous, and aerial bombardments are more appalling than most past atrocities. The radio, which physically is as marvelous as the airplane, propagates so many lies and insanities as to sicken its creators, notably the illustrious Branly. Talking movies, a marvel based on the photoelectric effect, are spiritually not much better.

Where is the "*moral* progress" in all this? This is what we do not clearly discern. Does not this expression denote one of those falsely created phantoms like "the underlying foundation of things" we questioned above?

Thus it does not appear certain that we have reason to count on moral progress to occur in historical time. And here too, shocking contradictions arise. A short while ago I was speaking of the esthetic value of science, of the appreciation of its beauty and of the keen pleasure that one derives from a labor of creation or discovery. Now I have often wondered if the technician who creates a new type of rifle or bomber does not experience a similar pleasure. If this is so—and I am afraid it is—the idea of moral progress arising from science cannot be defended.[2]

[2] Since this article was written, its general tenor seems to have been borne out by the facts. We now have the atomic bomb. Can its inventors take pleasure in its discovery?

To consider another order of ideas, stress must be laid on the reaction of biologists brought face to face with a geometrized universe. Such a universe is "humanly posited," and our thought is necessarily its substratum.

It may be that nothing good can come from "time," it is so imperfect a notion and so bound up with (astronomical) *matter*. What can morality have to do with it?

Morality seems *independent of time*, that is, *eternal* by definition.

Literary genius has sometimes expressed this in very gripping terms. Here is a sonnet whose beauty has something superhuman about it:

Le Cloitre[3]

Un crucifix de fer tend ses bras sur le seuil.
De larges remparts gris ceignent le cloître austère
Où viennent se briser tous les bruits de la terre
Comme des flots mourants aux angles d'un écueil.

Le saint lieu, clos à tout, gît comme un grand cercueil
Plein de silence, plein d'oubli, plein de mystère.
Des vierges dorment là leur sommeil volontaire
Et, sous le voile blanc, portent leur propre deuil.

Tous les ressorts humains se sont rompus en elles.
Dans l'éblouissement des choses éternelles,
Elles marchent sans voir, hors du Temps, hors du Lieu.

Elles vont, spectres froids, corps dont l'âme est ravie,
Etres inexistants qui s'abîment en Dieu,
Vivantes dans la mort, et mortes dans la vie.

Edmond Haraucourt

There is no difference between "universe" and "human conception of the universe."

Although a mathematician, I wonder whether the study of life is not the most fundamental of all problems.

[3] A prose translation follows:

The Cloister

An iron crucifix stretches its arms over the threshold. Wide gray walls encircle the austere cloister, against which all worldly din shatters like waves dying upon the edges of a reef. The holy place, closed off from everything, lies like a large coffin full of silence, full of forgetting, full of mystery. There virgins sleep their willing sleep and under their white veil wear their own mourning. All human springs of action within them have been broken. In the bedazzlement of things eternal, they walk without seeing, outside of Time, outside of Place. They go, chill specters, bodies bereft of souls, non-existent beings immersed in God, living though dead, and dead though alive.

Obviously the first tercet is what we should fix our attention on. The author there frees himself of time and space with more than Einsteinian mastery. This permits him to portray and idealize beings who are incontestably of a superior morality. The only thing is, they are virgins; the generalization of such a state cannot last. There is eternal life in this bedazzlement before eternal things; the search for perfection leads to annihilation.

We may add the further remark that if worship of the eternal came to be substituted for that of the temporal, progress could be expected at some definite epoch; the tendency toward the eternal would have a representation in historic time.

But the probability of such things happening is practically zero. Humanity seems to be possessed of a curse that is impossible to escape. Moral progress is more dependent on a certain mysticism than on improvements in science.

A better humanity would of course always live within a material framework and in a succession of days and years, but I am inclined to think that it would feel less imprisoned within it.

Perhaps we should view as progress the fact that belief in a unique truth is receding. Mathematics itself now has its uncertainties, even its contradictions. Fanatical insistence on the absolute truth of some particular opinion is losing ground. And the possibility of building along lines both esthetic and scientific seems to exist, at least in the realm of speculation. That is enough to justify the effort!

Toulouse, December 30, 1942.

42

THE NOTION OF GROUP AND THE ARTS

by Andreas Speiser

PROFESSOR AT THE UNIVERSITÉ DE BÂLE

THE Pythagoreans said, everything is number. Today we could both particularize and enlarge this thought and say, everything is group. In fact the concepts in accordance with which we see and shape the world are in the nature of a group. First of all, take space, which constitutes what we ordinarily call the exterior world and which gives reality to it. It forms a sphere whose center is everywhere and whose boundaries are nowhere. That is, it is unlimited and each of its points is a center of rotational symmetry. We could add that each plane in it is a plane of reflection and each straight line an axis of revolution. Now let us imagine a second example of this space moving in the first space, which is supposed rigid and at rest. A movement of the mobile space consists then in the set composed of an initial and a final position, however the transition is carried out. When the mobile space is in its final position, it can be set in motion by a new movement which carries it to a new position, the third, counting the initial position as the first. Evidently there exists a movement formed by the first and third positions; we call this the product of the two movements. The inverse of the first movement consists in the transition from the second position to the first. This is why we say that movements in space form a group.

Whenever we take possession of a world that is independent of us, we must apply concepts which have the character of a group. The reason is that we do not live in the instant but are creatures with a history and endowed with acquired faculties that ought to aid us independently of time and place. These faculties must be applicable at any time and place; therefore the concepts have, a priori, the character of a group. What space is to the exterior world, that is, both form and necessary condition, number is to the mental world. Plotinus calls these two concepts *pro-topos, protoposeis,* a word which is untranslatable unless one wants to perpetrate barbarism, like "pre-spacification." They establish the framework which gives order to great disorder. Any philosophy wishing to downgrade mathematics must first destroy the notion of order, as Bergson does; he speaks of disorder being equivalent to two orders [*désordre = deux ordres*] that is, he calls disorder an order of the same rank as order; it is only an unexpected or misunderstood order. But then, what difference is there, for example, between the noise of the instruments of an orchestra before the concert begins and the symphony which follows? Let us stop playing with words and continue to admire the order and beauty of the masterpieces of nature and art.

The numbers that constitute the interior world form a group obeying the law of addition. In fact, in the unlimited series of positive and negative numbers there is the possibility of translation. Every number is the center of symmetry for the set of numbers; it is surrounded in an identical manner by this set. Reality first reveals itself to us through ourselves; we know only the I, therefore the One. But how does one infuse life into the exterior world and how does one suppose other individuals? The notion of space certainly allows us to suppose another man as an exterior object, but how do we give him a soul and an I? To have the possibility of saying, "There is a man like myself," we must first have the concept which allows us to count him with ourselves and to say, "What he is with respect to me, I am with respect to him." Thus it is the principle of identical surroundings which permits us to project our I, that is, this One. Suppose that all this were only a dream—and the notion of space does not exclude this possibility at all—then our fellow-creature would be only my dream; and I myself, being in the same position with respect to him, would be only the dream of a dream. Now we know directly that we exist; thus it must be that the other person exists in the same way that I do. It is this syllogism which reveals to us the I of our fellow-creatures, and the child learns this through familiar proverbs from his earliest childhood.

The primitive mentality scarcely recognizes space and materiality. Everything exterior is projected through the force of number; thus this exterior world consists only of subjects, of living and spiritual beings. The difference between things actually experienced and things dreamed of hardly exists. Thus, one morning the chief of an African tribe recounts his dream of a long voyage to Europe, and his subjects listen with astonishment and congratulate him on his great success and admire his courage. The notion of death as the end of everything is still unknown. To the extent that the dead man is remembered, he exists almost as if he were still alive, and when he has been forgotten, one cannot even state that he is no longer there. Even material objects receive life through the force of number. These are the I's which have a magic, personal power upon the surroundings. The trees, rivers and mountains live and take a human form which appears before privileged mortals. The notion of space which permits us to deal only with lifeless objects has destroyed this magic for us to such an extent that we are no longer capable of invoking the charm of the primitive world.

The two groups of numbers and of space thus preform the exterior world; the former permits us to endow it with life, the second reveals its lifeless objects. The purpose of art in general is to infuse the exterior world with life; it is thus a priori probable that it employs these two groups in the sense that it regards the numbers as a continuum. It succeeds in this undertaking by means of symmetries, and I would like to demonstrate the role that the notion of group plays here. But in order to do so I must deal more specifically with this concept.

The theory of groups starts with a simple definition: A group consists of a finite or infinite collection of elements which obeys the associative rule, for which an identity element exists, and each element of which has an inverse element. The goal of the theory is to discover all the groups which exist and give their principal properties. First of all we must know what a property of a group is. We are still far from a complete solution to this problem. However, for finite groups we can say that we have reached a certain knowledge concerning the phenomena that arise. As to the history of these discoveries, several periods can be distinguished. The first begins with Euler, who described a certain number of groups in his geometry and theory of numbers. Lagrange, Gauss and Ruffini followed. The second period commences with the Norwegian mathematician Abel and the Frenchman Galois, to whom we must add Cauchy. Camille Jordan's treatise on substitutions, appearing in 1870, forms the magnificent apex of this period. Following Riemann's work, there are Schwarz, Poincaré, Klein and Lie. These men extended the notion of group to the theory of functions, differential equations and differential geometry where a fundamental connection with non-Euclidean geometry, which had heretofore been considered a purely formal concept, was discovered. Starting with these discoveries, groups have taken over all of physics, where they now play the role of a universal "logos." This triumphant conquest is one of the greatest marvels of the history of science. Since Euler's solitary excursions into the domain of arithmetic, it has taken but two centuries to make this terrain accessible to anybody.

Coming back to the subject of art, I would like first of all to quote the great mathematician Henri Poincaré. He states (*Science et méthode*, Paris, 1908, p. 57):

> One may be surprised to see what emotions are evoked by certain mathematical proofs seemingly of interest only to the mind. This would be to forget our feeling for mathematical beauty, for the harmony of numbers and forms and for geometric elegance. This is a true esthetic feeling which all true mathematicians know. And that is true sensitivity.

And Proclus says in his commentary on Euclid, "Wherever there is number, there is beauty." I believe that one can invert this latter phrase and say, "Wherever there is beauty, there is number." In fact, it is through number that we impart life to things; now art has precisely this goal; thus it brings the power of number into play. In order to demonstrate the role of groups and numbers, I would like to give some examples. Egyptian ornaments such as those found in the tombs at Thebes form spirals which repeat indefinitely and cover the surface. It has been calculated that there are seventeen different possibilities of symmetry of this type for a plane surface, and the majority of these were found in prehistoric times. From these purely mathematical and mystical configurations arose the half-geometric, half-vegetal ornaments of Minoan art. The Greeks, perhaps Archimedes himself, added geometric mosaics. They still play an enormous role in decoration today, to the point where it could be said in 1893 (Flinders Petrie, *Egyptian Decorative Art*):

> In practice it is very difficult or even impossible to pick out a decoration which can be proved to be of independent origin and not copied from the Egyptian stock.

It is true that from 1900 on there has been an attempt to construct new designs without the aid of mathematics, but these are real horrors which are usually allowed to disappear after a brief period.

We should add the polygons of the Arabians to the Greek mosaics. They consist of a single line which repeats and twists, forming various figures by means of its interweavings, such as stars, squares and even pentagons. With the aid of color the same ornament can assume a considerable number of different aspects. Generally these are sub-groups which have the same coloring, and it is by this means that the forces of symmetry become effective. In a properly designed ornament each element of symmetry must be made evident by a special mark.

The same principle of identical polygons is employed in constructing lace on bobbins, but it is only in music that its inexhaustible power is revealed. In fact a canon is only a polygon in sound which when set against itself produces a certain number of measures full of geometry. Composers probably proceed visually. In listening to a canon, our mind must follow each voice separately and at the same time pay attention to the vertical harmony. It requires difficult and persistent work to learn to fall under the spell produced by this sublime art. This sort of music is often considered a more or less sterile game, an opinion contradicted by the fact that the greatest composers, among them Bach, Mozart and Beethoven, have devoted a large part of their efforts to the construction of canons and fugues, in which they acquired an astonishing virtuosity. Canons were heard in the taverns frequented by Falstaff, and they can be heard today at concerts of modern music; they form both the heart and mind of all music. Just as, according to Poincaré, intelligence and feeling form an inseparable unity in mathematics, so is it impossible to distinguish heart and mind in music, whatever Blaise Pascal may say.

I could continue and show how all melodies and all harmony are impregnated with number and with geometry and how proportion gives life to painting and to lyric poetry, but I shall stop here and merely give the opinion of the great composer Rameau. In his treatise on music, he says that it is not music which is part of mathematics, but on the contrary, science which is a part of music, for it is founded on proportion, and the resonance of a sonorous body generates all the proportions. D'Alembert was wrong to laugh at this phrase, for it contains a great truth. Wherever there is number, there is beauty, and we are in the immediate vicinity of art.

43

ARCHITECTURE AND
THE MATHEMATICAL SPIRIT

by Le Corbusier

We are inclined to believe that the literature and arts of our time have ignored two aspects of the civilizing function of mathematics. They have sacrificed the rigor which represents the part that clear consciousness plays in anything creative; and they have ignored one of the most original sources of lyricism.

OUR editor, Le Lionnais, addressed these lines to the authors from whom he solicited contributions to this book.

Here are some additional lines excerpted from a contribution of mine to a forthcoming issue of *L'Architecture d'aujourd'hui* to be devoted to the *synthesis of the major arts*:[1]

To take possession of space is the first act of living things, men and beasts, plants and clouds; it is a fundamental manifestation of equilibrium and of duration. The first proof of existence is the occupation of space.

The flower, the plant, the tree, the mountain are upright, living in an environment. If one day they attract attention by their truly reassuring and sovereign attitude, it is because they appear limited by their shape but induce resonances all around them. We stop, realizing that there is so much natural interrelationship; we look and are stirred by so much concordance orchestrating so much space, and we measure while what we observe is irradiating.

Architecture, sculpture and painting are specifically dependent upon space; they each have to manage space by appropriate means.

---

[1] An appeal bearing this title was inserted in my first page of the weekly *Volontés* at the end of 1944.

The essential point I will make here is that the key to esthetic feeling is a concern with space.

Action of the work of art (architecture, statue or painting) upon the surroundings; waves, shouts or outcries (the Parthenon on the Acropolis in Athens), beams shooting forth like those caused by radiation or an explosion; the land near or distant is shaken by it, affected, dominated or caressed by it. Reaction of the environment: the walls of the room, its dimensions; the public square with its façades of varying weights; the sweep of the land or its slopes, and even the bare horizons of the plain or the rugged horizons of mountains: the whole environment imposes its weight upon this place possessing a work of art, this mark of man's will, and imposes upon it its depths or its protrusions, its hard or fuzzy densities, its violence or its tenderness. We are presented with a phenomenon of concordance, exact as mathematics and a true manifestation of plastic acoustics; thus we are allowed to invoke one of the most subtle orders of phenomena, the bearer of joy (music) or of oppression (cacophony).

Without making the slightest special claim, I shall make a statement about the "magnification" of space that artists of my generation ventured upon around 1910 under the influence of the prodigiously creative spirit of cubism. They spoke of a *fourth dimension*, whether more or less intuitively and clairvoyantly does not matter. A life devoted to art and especially to the search for harmony, has permitted me, through the practice of the three arts of architecture, sculpture and painting, to observe the phenomenon in my turn.

The fourth dimension seems to be the moment of complete escape, brought about and triggered by an exceptionally close harmony among the plastic means employed.

This is not due to the theme chosen but is a victory of proportion in all physical aspects of the work as well as for the efficiency with which intentions are carried out—intentions which may or may not be under control and apprehended, but which do exist and are due to intuition, that miraculous catalyst of acquired, assimilated and even forgotten knowledge. For in a work that is completed successfully there lie buried numerous intentions, a veritable world of them, revealing themselves to those entitled to them, that is, those worthy of them.

Then a bottomless well opens up, wipes out walls, drives contingent presences away and *accomplishes the miracle of ineffable space.*

I do not know about the miracle of faith, but I often live the miracle of ineffable space, the crowning of plastic emotion.

We are quite alive to the fact that the precision required in all these acts meant to trigger a superior emotion is of a mathematical order. One word expresses the product: harmony. Harmony is the happy co-existence of things. Coexistence implies a double or multiple presence; consequently it calls for relationships and accords. What kind of accords could these be, to interest us? Accords between us and our environment, between man's spirit and the spirit of the things, be-tween the mathematics which is a human discovery and the mathe-matics which is the secret of the world.

This can lead to trances or religious transports. At the other extreme, simple and solid, is our daily work as artists. The reality and ingenuity of precise, measured and measurable relationships go into this work which fills the plastic artist's day. The plastic artist will be a poet, that is, able to stir the emotions, only when his materials are of unquestion-able quality, made with the unrelenting rigor of striking relationships.

Unrelenting rigor of striking relationships. . . . Allow me to cut short the insatiable dialectic of words in order to explain some material, experimental facts.

I

THE RIGHT ANGLE

In 1923 I built a small house. The grounds were minuscule. There was a low retaining wall at the edge of the lake (Figure 1). When the

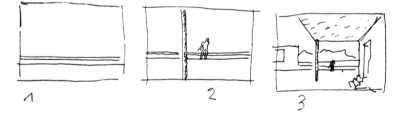

house was finished, a plastic phenomenon compelled attention: the crossing of the wall at right angles by a narrow column consisting of a simple metal pipe (Figure 2). The scanty grounds were deliberately enclosed within walls to mask the horizon. A view of the lake and mountains was permitted at just one specific spot where means of measurement existed that could make one feel strong relationships:

the intersection of the right angle, which can be sublime; the back-lighting and the luminous expanse of water; the complete sobriety of the architectural lines and the sharply engraved outline of the mountains (Figure 3).

Man and his measuring tool: the intersection of the right angle. Nature and her discourse: space, the horizontal plane of the water, the narrative told by geographic and geological profiles. . . .

Two alternatives presented themselves:

the scribble or the orthogonal.

II

The Need for Mathematical Expression

I built my first house at the age of 17; it was decorated from top to bottom. I was 24 when I did my second; it was white and bare: I had traveled. It is 1911, and the plans for this second house are on the drawing board. The arbitrariness of the holes in the façade (the windows) becomes startlingly obvious. I blacken them with charcoal. At once the black spots speak a language, but this language is incoherent. The absence of rules and laws is obvious. I am overwhelmed: I realize that I am working in complete chaos. Here is when I discover the necessity for the intervention of mathematics, the need for a monitor. From now on this obsession will occupy a corner of my brain.

III

Memories are Awakened

I am an apprentice engraver of watch cases. It is around 1900; decoration inspired by natural elements is in style. I am in the mountains, drawing an old fir tree that stands in the pasture. I discover a law. "Look," I say to my master," you can tell the age of the tree from its oldest branch."

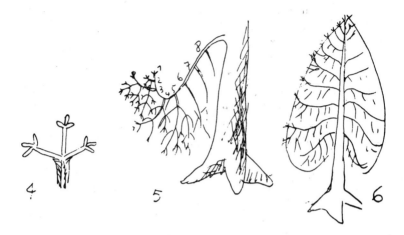

Here are the three growths of the year, each with its three buds (Figure 4); (a) will provide next year's growth; (b) and (c) will angle off, each in turn yielding three growths with three buds apiece. The law is enunciated. The oldest branch, the one closest to the ground, almost surely initiates a series of growths (Figure 5).

And the entire tree (Figure 6) is a pure mathematical function. (This is not an assertion of fact; I have never had the opportunity to prove it.)

IV

Equilibrium Through Equivalence

This same year the cook in the Alpine inn where I am spending my vacation has served chamois. I come down to see him, anxious to carry off the waste trimming of the head—the horns. Too late! "Take the hoofs at least!" I remained several days skinning the chamois hoofs with my pocket knife, and musing on their slenderness, as well as on the formidable number of ligaments joining the bones, insuring articulation and attachment to the muscles.

The chamois leaps from rock to rock, its entire powerful body coming to rest on four small supports (Figure 7).

A day arrived later when I was studying about reinforced cement, in books and on the job simultaneously.

Another day came, later still, when I was meditating on a series of concomitant facts (urbanism, the esthetics of reinforced concrete,

economy, the joyous exploitation of the dazzling resources of the new
material), and I arrived at some stimulating, though scandalously

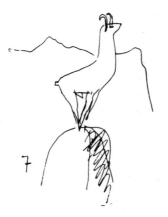

heterodox, conclusions concerning plastic form and doctrine; free-
standing skeletal structure, glass faces, stilts, etc. in violent contrast to
architectural traditions, usages and attitudes, namely: stilts (Figure
8), free bearing-structure and free façade (Figure 9), in opposition to
the previous harmonies springing from stone and wood (Figure 10).

The hoof of the chamois and the stilts of reinforced cement take into
account the material that goes into their making, creating balanced
wholes that are intense, nimble and intelligent. Ideas to enchant the
mind and overwhelm the conformists; and finally to repudiate
Vignola!

V

ALL THINGS ARE SUBJECT TO,
PROCLAIM OR DEMAND A LAW

Thus are organisms created and shaped, guided to their form by evolution and selection.

Let us suppose a dwelling or office building, in short a shelter for man.

In 1922 (at the Salon d'Automne, "A contemporary city of three million inhabitants"), there was an exhibit of a "Cartesian skyscraper" which presented a contrast to the arbitrary forms, dimensions and ground plans of the American skyscraper. The law here was to meet the requirements for light, something which had preoccupied me from the very beginning.

This was the *cruciform* skyscraper, a veritable radiator of light (Figure 11).

But in resuming these studies around 1930, I estimated that at least a quarter of the façades would thus have a northern exposure, and this appeared absolutely atrocious to me. Consequently the form of the skyscraper undergoes a change:

The sun's course is represented by the two solstices (Figure 12).

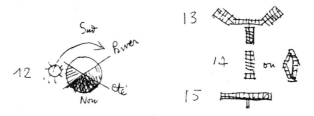

Illustrated is a desirable form for the building, the sun striking it at the east, south and west (Figure 13). Another in the shape of a thorn, with the sun striking from the east and west (Figure 14). Another form, the whole facing front, with the sun at the south (Figure 15). These are not the whole story.

According to studies as well as experience, the law of the sun imposes new arrangements:

The sun of the dog days coming through glass façades is over-powering; yet glass façades are a masterly gift of modern technology, not to be rejected. Certainly at the south and west, and probably at the east, we must "break" the summer sun. But in winter the solar rays must be able to penetrate the building. The summer trajectory is high and the brise-soleil (sun-control louvre; Figure 17) in front of the

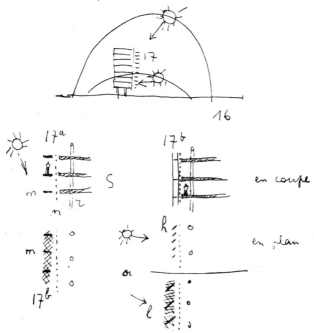

m = horizontal brise-soleil; n = glass façade; r = bearing stanchions; S = floors.

glass façade will shade the glass while permitting diffuse light to enter. The low winter trajectory is not affected by the summer brise-soleil, and direct sunlight penetrates the interior.

At the east or south the sun is harmful only in the later hours of the day when infrared waves are present. A brise-soleil placed in a pre-dominantly horizontal position will be sufficient (m in Figure 17).

To the west, the sun's heat is at a maximum at sunset, that is, when it is near the horizon. The brise-soleil will be predominantly vertical:

h with relatively narrow and close vertical slats; *b* with wide vertical slats spaced farther apart.

We have here demonstrated an increasingly precise biology; buildings to be used for business or dwellings are to adapt their form in accordance with the law of the sun, the master of life. A law will govern the very essence of architectural substance.

VI

HARMONY IS AN ACT OF CONSONANCE

If, in order to house 100,000 inhabitants of Algiers, I once proposed occupying the site called "The Heights of Algiers" (which hangs precipitously over the harbor 400 meters as the crow flies from the center of town) with volumes of construction which aroused enthusiasm as well as indignation throughout the world, it was because I was

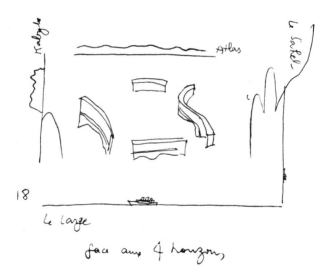

dominated by the imperative and deep-seated need to enter into a harmonious relationship with the surrounding universe, that is, with the site, the sun and the topography. These considerations gave rise to forms conditioned by, and wholly responsive to, this triple objective. This is what might be called *situating* something: harmonizing a human work with its environment, bringing man's mind into accord with nature's laws. To make them resound, make them sound good together, to produce consonance. To establish the reign of harmony. To accomplish, if successful, the miracle of ineffable space.

VII

Ineffable space, produced by mathematical rigor applied to the whole complex of problems one faces. The senses transmit to the mind the action of the forms which determine the extent of the sensation. To appreciate that the measure of the latter has absolutely no relation to size, we need only consider the humblest sea shell. Human enterprise can, in turn, embody in the purity and harmonious arrangement of the shell a vast number of concordant elements and events, and thus achieve radiance.

The hall of the Palace of Soviets* (1931) was to hold 14,000 spectators and auditors. One possible source of conflict was eliminated at the outset; the problem of statics was separated from visual and auditory problems. All three could have been combined, but incompatibilities in the systems would have rendered the attempt hazardous. Bearing and supporting members obey laws of gravity and statics and are subservient to materials and construction processes. The phenomenon of acoustics is something else again; sound waves ricochet off reflecting surfaces, and the intensity with which they are transmitted is a function of the distance covered. We are no longer dealing with a static order but rather with a biological order—mouth or ear, emitter, transmitter and reverberator. Finally the visual phenomenon in the very difficult conditions considered here, where distance chops up the spectacle, prompts us to seize upon whatever advantages we can. We shall see how.

The listener-viewers were installed in a concave amphitheater permitting everyone to have a good view of the stage but also giving a real view of the mass in attendance. For if the amphitheater had been required only to provide visibility of the stage, the seats would have been arranged in horizontal rows, and the plane thus determined by the spectators would become tangent to their line of sight, and the view of the crowd itself would be reduced to practically nothing. The concavity of the amphitheater, on the contrary, provided a display of the occupied surfaces, and each spectator could see with his own eyes

---

* Le Corbusier entered this project in a competition but did not win [translator's note].

the tremendous crowd intent upon the action on the stage. This is in itself a powerful and moving source of emotion.

The techniques of sound led us to a solution of great purity. The direct sounds—speakers, singers, orchestra and all other sources (J in Fig. 19)—were picked up by microphones (Fig. 19, H) placed above the stage. The stage walls were made absorbent: A "gulf" of eleven meters was created between the first sound sources and the first listeners. Then, the undistorted sound was entrusted to a single loud-speaker (Fig. 19, F) placed in front of the stage, which thus became

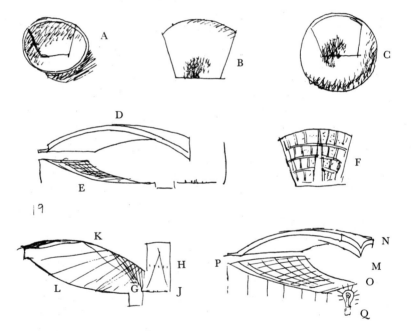

the only emitter. The problem was being simplified: carry this sound emitted from a single point to 14,000 pairs of ears (Fig. 19, O-P). All that would be needed thereafter would be to establish the necessary reflecting surfaces, taking into account a loss of sound proportional to the square of the distance; consequently, to construct a reflecting membrane (Fig. 19, K) mathematically proportioned to take into account the location of every listener present and his distance from the loudspeaker. The ceiling of the hall would satisfy this requirement.

First of all choose a concave form for the amphitheater; here is a fragment cut out of a basin (Fig. 19, A).

Next, agree that the seat of every listener (Fig. 19, L) will be sprinkled with droplets of sound emanating from a mouth (the loudspeaker; Fig. 19, G) and reflected mathematically by the ceiling (Fig. 19, K), with the direction and intensity of the reflection entering into the mathematics. Where shall we get a serviceable form (Fig. 19, B). In a torus whose inner circle is reduced to a point (Fig. 19, C). The farther along the reflecting ceiling the droplets of sound strike, the more numerous they should be, and as a result the more extensive the ceiling surface will have to be. In this way good acoustics are achieved; the seats at O and the seats at P in the figure enjoy practically the same harvest of sound waves. This was verified in a mock-up through the use of electric lighting. A small lamp was placed at Q. From O to P the amphitheater was illuminated evenly by reflection from the ceiling, with the same intensity at every point. A phenomenon of perfect concordance.

Elements of the solution became apparent: the 14,000 seats (E); the reflecting shell, an acoustic phenomenon (D). The amphitheater (L) and stage (J), a visual phenomenon. At this point I have a question: in principle I am hostile to gigantic programs; they lose sight of the human scale. If, by precise methods, sound can be transmitted infinitely farther than was imagined, there are still severe limits to visibility. My question then is as follows: To look for a needle in a haystack or a pile of dirt is unrealistic. But finding a needle becomes possible in a place where all points—the floor of the amphitheater, the lateral partitions, the back, each point of the vault—are in mathematical agreement, establishing a close and harmonious relationship between the immense hall and the stage. The needle (in this instance, the actors on the stage) becomes the fateful target. I shall answer my own question. We have to admit that a strong feeling of unity and coordination comes into play; the senses are swept towards a single objective, and the result is that spectators, listeners and actors find themselves united in the same unique and intense adventure.

The hall has become a place of harmony.

One readily understands that without bold initiative it would have been difficult to counterpoint the matter with considerations of statics and the resistance of materials.

The ceiling (the reflecting shell) can be made of a slab of reinforced cement from three to five centimeters thick. At a distance of two meters from this membrane there will be a second slab serving as an umbrella. A scheme of very delicate, trellised small beams will serve to join and solidify the two conches. Various results are achieved: acoustic quality of the lower slab, impermeability of the upper slab, possibility of inspection and repair, lightness, etc.

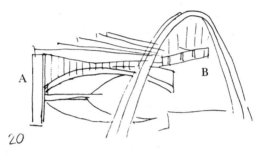

At this point it is only necessary to suspend this double keel from above. Support on the ground: eight pylons at the back of the hall (Fig. 20, A). Another support: a parabolic arch of reinforced concrete 100 meters high, whose two feet rest outside the building. Hanging from king posts is a beam, a sort of binding joist (B). At the top of the pylons (A) and on this latter joist (B) rest eight large latticed girders of reinforced concrete, fitted with king posts from which the vault will hang. This immense ceiling, like Holofernes' head, will be held by its hair! So that the phenomenon of "the law of gravity and methods of construction" will be in evidence outside the building itself, *on the exterior*. The interior is like a pure sea shell. And save for errors, we are adhering to the most authentic tradition of cathedrals except for the fact that we are dealing with reinforced cement whereas the Gothic builders used dressed stone without binding material.

We cannot bring this study of the hall of the Palace of Soviets to a more fitting conclusion than by repeating once again the remark so pointedly phrased by Le Lionnais:

> . . . The arts of our time have ignored two aspects of the civilizing function of mathematics. They have sacrificed the rigor which

represents the part that clear consciousness plays in anything creative; and they have ignored one of the most original sources of
lyricism.

To the artist, mathematics is not just the subject matter of mathematics. Mathematics is not a question of calculation perforce but
rather the presence of royalty: a law of infinite resonance, consonance
and order. Its rigor is such that truly a work of art results, be it a
drawing by Leonardo, the startling exactitude of the Parthenon, the
cutting of whose marble can be compared with the work done by a
machine tool, the implacable and impeccable construction of a
cathedral, the unity which Cezanne achieves, the law determining
a tree, the unifying splendor of the roots, trunk, branches, leaves,
flowers and fruit. There is nothing haphazard in nature. If one has
understood what mathematics is in the philosophical sense, all of the
works of nature will henceforth reveal it. Rigor, exactitude are the
means to a solution, the source of character, the reason for the harmony. "The lion is recognized by the blow of its paw."
The mark of measure. Giving a measurement, taking a measurement, making measurement reign: these acts are necessary to establish
order and the means of order.
The traditional measurements of a finger's breadth, inch,* foot and
cubit are expressions of natural mathematics. A great crime was
committed against the domain of human construction when the
metric unit was imposed upon it, that ridiculous and pointless forty-
millionth of the earth's meridian!
Dimensions and measure lead to the establishment of proportions
in things. Laws administer the proportions which produce character.
In the domain of construction attention to proportion has, since the
Renaissance, declined to the point of disappearance. I feel the idea of
proportion as part of me, and my mind as well as my hand never
cease to be occupied with it. In architecture sketches are guides; in
painting, too, sketches are guides. You can attain such mastery in this
mathematics of the plastic arts that you are no longer obliged to make
calculations and preliminary drawings; your hand will perform them
automatically. It remains for our modern world to free itself from the
arbitrary metric measurement in construction and replace it with the

* In French, *pouce* means "inch" and "thumb," and *coudée*, like "cubit," is
from the Latin word for "elbow" [translator's note].

tremendous resources of numbers, particularly the fruitful and in-exhaustible golden section. . . .

I have been dreaming for a long time of a mathematical unity resulting from a golden rule applied to projects dealing with urban problems, architecture and furnishings in our machine civilization.

After twenty years of study, I believe that I have lighted upon, not the rule, but one of the rules capable of triggering a prodigious flowering and abundance of harmonious forms. In the United States as well as in France, in the U.S.S.R., in England and everywhere else, this rule could serve as a tool for determining the dimensions of any prefabrication or any construction. I am announcing this discovery here to whet the curiosity of the seekers and so that this rule, investigated and perfected by all, may before too long aid in projecting unity and harmony into the projects of this second era of machine civilization now underway. Deep-seated and universal harmony, part of our epoch's potential, charged with the mission of eradicating the chaos in which the machine civilization was born.

> January 4, 1946, in the Azores,
> aboard the cargo boat *Vernon S. Hood*.

44

MATHEMATICS AND MUSIC

by Henri Martin

PRINCIPAL INSPECTOR OF TECHNICAL INSTRUCTION

LEIBNIZ once wrote:

> Music is a secret arithmetical exercise and the person who indulges in it does not realize that he is manipulating numbers.

He could have added, and a person playing the harpsichord does not realize that he is manipulating logarithms. Indeed the connection between music and certain parts of mathematics is very close, primarily for the following reasons:

(1) The effect of a musical sound upon our ears depends first and foremost upon its pitch (physicists speak of its "frequency," i.e., the number of vibrations per second of the body emitting the sound). To say that we hear middle c, or to say that our ear registers 256 vibrations per second, means the same thing. Therefore a number is associated with each sound, and conversely, with each number, whether integer or not, is associated a sound.

(2) Hearing two sounds simultaneously is equivalent to perceiving two numbers and a relationship between them. To hear the first and fifth tones of the same scale is equivalent to "hearing" the ratio 3/2, which is the ratio of their frequencies. Now experience with music shows that the esthetic effect of a chord depends almost exclusively on the ratio of frequencies; the whole question of harmony is therefore a question of the choice of ratios.

(3) To these two reasons we add rhythm, which is by its nature essentially arithmetical. We shall put this question aside because it is at once too important to be mixed in with another and too removed from the study of sounds which we propose to undertake.

Let us now note that colors, which are also differentiated by their frequencies, play the same role with respect to sight that sounds play with respect to hearing. Our eyes and our ears are *counters of frequencies*. However, there is an enormous difference between the way colors and sounds are used: A painter can put colors of any frequency whatever on his canvas, while a composer cannot place sounds of arbitrary pitch in his work. Why is this?

First of all, because he must write his music. There would have to be an infinite number of symbols to designate all the pitches; deciphering such notation would be almost impossible and very slow in any case. Next, music is made to be played, and the large majority of our instruments can produce only a limited number of sounds.

Furthermore, our ear is incapable of discerning two sounds that are too close. This "power of separation" obviously varies a great deal among individuals, but it scarcely permits one to distinguish a shift of the fingers on the violin of the order of two millimeters, for example. This therefore makes it fruitless in any case to use all the frequencies. However, it is agreed that a practiced ear can distinguish about 300 sounds in one octave; this is still too much for musical notation and for the capacity of the instruments (a piano of eight octaves would have 2400 keys).

Therefore, for musical purposes, we are led to employ only a limited number of sounds in each octave, which is the natural basic interval. Let us recall that two notes are said to be an octave apart if the frequency of one is double the frequency of the other; or, what comes down to the same thing, if a string produces a certain note, half the length of string will produce the octave.

The following question was asked as soon as music became a social art: Among the 300 discernible sounds in an octave, how does one choose the scale of sounds to use?

We would like the reader to grasp the importance of this question; in a way, it committed music for millennia, if not for eternity.

Once a scale was adopted, it became in fact practically impossible to change it. Let us take an example. Suppose that the musicians of a certain era had decided to divide the octave into 10 intervals in the following fashion: The fundamental sound is given by a string one meter long, then its octave by a string 50 centimeters long; the intermediate sounds will be given by strings of 55, 60, 65, etc. centimeters, with five-centimeter intervals. A priori there is nothing disturbing in this. Ah, well! A piece written in this scale cannot be transcribed with our present notation, even with the aid of sharps, flats, double sharps or double flats. In any event, only certain instruments, the violin among them, could perform such a piece.

The music of such an era would be completely unavailable to us, therefore "dead."

The continuity of musical life thus requires, assuming that conditions pertaining to the instruments be such as have prevailed up to now, the continuing use of the same scale or those with practically negligible differences—which in fact is what has happened in the course of the history of our music.

Were musicians and philosophers to reflect about all this, they would be alarmed at the responsibility assumed by the first musical theorists when they cut up the octave into definite units. In no area of art did a decision have such importance; it is perhaps the Greeks' most splendid and most eternal claim to glory that they created the scale *at the same time that they created mathematics* (the coincidence is worth noting).

If another scale, audibly different, had later been revealed to be more esthetic, efforts would have been made to establish it, and despite the difficulties we have indicated, the centuries would have won out over a division of the octave judged outworn. Now, nearly 2500 years have passed and the present-day scales, which we shall investigate, are in fact only variations of the Greek scale; a single notation serves them all; their notes have the same name, and a piece is composed, written down, played and sung without specification of the scale.

If they are not entirely equivalent physically, they are used and regarded as such.

From the point of view of philosophy and esthetics, the question of the timeless value of the Greek scale is accordingly disturbing! Let us try to clarify this mystery partially; to do this we must first specify the nature of the three most frequently employed scales: the diatonic Pythagorean scale, the Zarlin or physicists' scale and the tempered scale immortalized by J. S. Bach.

We shall frequently speak of the interval determined by two notes; we should understand by this the ratio of the frequencies of these two notes. Take, for example, the notes corresponding to 400, 600 and 800 vibrations per second; the interval of the first two is 600 to 400 or 3/2; the interval of the latter two is 800 to 600, or 4/3.

The difference between the frequencies is the same, but the intervals are not equal; we must not lose sight of this concept in what follows.

1. The Greek Scale

Let us take a string which produces the sound *f* and let us regard this as the beginning of an octave. Two-thirds of this string's length will produce a higher-pitched note, which by definition will be called the fifth above *f*: This will be our *c* of the same octave. Two-thirds of the *c* string will again produce a new fifth, the *g* of the octave immediately above; by doubling this *g* string we shall come back to the *g* of the initial octave, and so forth from fifth to fifth. The notes obtained will be in the order: *f–c–g–d–a–e–b* when transposed to the original octave, they appear in the order: *c–d–e–f–g–a–b–c*. Continuing this series of fifths beyond *b* does not give us *f* again, but a note called *f*-sharp, then *c*-sharp, etc., and when the series is continued on the other side of the initial *f*, it gives flats.

The concept of this scale is therefore extremely simple and coherent, and well within the Pythagorean tradition.

2. The Zarlino or Physicists' Scale

The principle here is entirely different. It consists of the a priori assumption that two sounds will be the more agreeable to the ear, especially if heard simultaneously, to the extent that they offer harmonics in common. Let us recall that an initial sound has harmonics, that is, sounds which correspond to doubled, tripled and quadrupled frequencies, etc., and are thus called second, third, fourth, etc., harmonics.

Let us take, for example, the frequencies 400 and 500 which correspond to an interval of 5/4; the fifth harmonic of the first sound will coincide with the fourth harmonic of the second (i.e., 2000 vibrations per second). Their harmonics, 10 and 8 respectively, will again coincide, etc. Complex intervals thus correspond to coinciding higher harmonics, and if the frequencies are incommensurable, the two sounds will not have harmonics in common.

A rather small number of intervals are thus determined. Among these we find some that figure in the Greek scale (9/8 for the interval of the second (*c–d*), 3/4 for the fourth (*c–f*), 3/2 for the fifth (*c–g*)).

Others do not figure in it but are close enough to Pythagorean intervals to be substituted for them and receive the same name.

The interval of the third (*c–e*) in the two scales is not the same, but the difference is practically unnoticeable.

Multiplying its frequency by 25/24 sharps a note; multiplying by 24/25 flats it; this produces sharps and flats that are very close to the corresponding sounds of the Greek scale.

3. The Tempered Scale

It must be added that the two preceding scales, by lumping together the very close notes *c*-sharp and *d*-flat, etc., defined twelve slightly unequal intervals. The tempered scale also divides the octave into twelve intervals, but makes them equal a priori.

As a result of this the twelfth power of each of these intervals is equal to 2, the interval of the octave. In other words, the fundamental interval is the twelfth root of 2, and the frequencies of the twelve notes are in geometric progression. If the first note is *c*, the second will be called both *c*-sharp and *d*-flat, etc., and will be the identical note in this scale.

The simplest way to determine the fundamental interval, the twelfth root of 2, is to make use of logarithms. This fundamental interval is an irrational number; as a result, the tempered scale *does not possess any simple interval*, a fact which would have driven Pythagoras to despair; and the notes of which it is composed *have no harmonics in common*, which is very far from the physicists' conception of the affinity of sounds. Nevertheless, the notes of this scale are close enough to those of the two preceding scales to receive the same names; but none of them coincides exactly with its homonym.

Thus the tempered scale is clearly based upon a more complicated mathematical conception than the others and could not have been conceived before the invention of logarithms.

Johann Sebastian Bach, who was the first to employ the "tempered scale," was able to do so only because Napier had invented logarithms before him, shortly after 1600. This scale has become, naturally, the one in accordance with which instruments of fixed pitch are tuned.

Other scales in addition to these three have been conceived. Fanatic partisans of the golden section have tried to apply it to the lengths of strings. These attempts, although very interesting, have not dethroned the standard scales, which remain the only ones in use;

the plural "scales" is, as we have seen, to all intents and purposes a singular, and Bach's scale is in fact merely the daughter of the Greek scale, from which it is not distinguished in practice.

It remains for us to attempt a mathematical-philosophical critique of these three scales to see if it would not be possible for mathematics to play a role in reconstituting the art of sound.

Their esthetic value is not to be disputed; it is an established fact by virtue of the emotional power of music. So we ask the following questions:

(1) Where does this esthetic value come from?

(2) Do these scales exhaust the possibilities of musical expression?

(3) Would it be possible and interesting to use other sounds than those which they impose on us?

The Pythagorean and Zarlino scales are, as far as mathematicians are concerned, identical conceptions; they reduce to simple ratios. The tempered scale, which uses irrational ratios, can only account for its emotional value with the feeble argument that its intervals are equal. In that case its creators could have been asked: "Why twelve intervals in the octave, rather than seven or fifteen, etc?" The answer seems obvious to me: "Because, by the division into twelve, we find ourselves with sounds that are practically identical with those of the Greek scale." The tempered scale was thus created to resemble the Greek scale. The same remark goes for the Zarlino scale, which chooses only those sounds already defined by Pythagoras, from among all those available to it.

The problem of the esthetics of the scales comes down entirely to a consideration of the Greek scale.

Its artistic value resides in the following concept, which has already been found to be true for the arts in general: Simple relationships create beauty. Moreover, it is *a fact*, established by the existence of works of art that have been conceived in accordance with this principle. Therefore we can state (with all the caution that should accompany such assertions): Our scales are beautiful because they partake of the Greek principle of harmony, the use of simple ratios.

Here the mathematician at once asks us whether the converse of this assertion also holds: "Can beauty be created only by simple ratios?"

The answer is no. Let us examine from a musical point of view some facts of a personal or general nature which will bear out this assertion.

I remember a climb in the Alps when there suddenly came to us from both slopes of the mountain the sound of cow-bells belonging to two herds, a sound which could scarcely be made out in the valley some hundred meters below. The great variety of the sounds which reached us constituted an astonishing euphony, without rhythm or classical harmony. I am unable to remember this without strong emotion. And yet, it consisted of sounds that cannot be transcribed in our music; no composer could have put them down on paper, and no orchestra could have attempted to recreate them. I venture to add that even an uncultured, but normally constituted, human being could not have remained unmoved by this music.

And the song of birds? It is a cliché, but true nevertheless, that the long phrases of the nightingale's song are beautiful; yet it is not possible to write them down in our scale.

In his *Voyage to the Congo*, Gide spoke of Negro melodies; he affirmed that the sounds employed in them do not correspond to those of our scale and that the transcriptions he attempted with our symbols were only approximate (he said so himself). However, these melodies stir the Negroes; therefore they are beautiful. It is true that this music eludes us (2500 years of musical atavism cannot so easily be uprooted!) but we do not have the right to deduce from this that it is deficient in esthetics.

As concerns another category of ideas, our scales and the majority of our instruments do not allow the use of sounds with continuous pitch variation, examples of which are modulations of the wind in the trees and under a door, or, in the area of rather disagreeable sounds, a siren.

As a result of all this our music does not and cannot exhaust the resources of the art of sound. Let us not conclude that our musical art is tainted with inherent poverty; there are more than 400 million ways of grouping the twelve notes of the scale (not taking into account possible repetitions or variations in rhythm); composers therefore have plenty of elbow room.

However, nothing prevents us from conceiving a new utilization of sound. To make ourselves clear, we are not considering here the creation of a new scale that will dethrone the one to which all our musical tradition has tied us much too much. We are concerned with supplementing the present possibilities with other possibilities in which mathematics could play its role.

Nothing prevents a musician from asking a mathematician to

design for him, for example, a scale of equal intervals, but with four-teen notes instead of twelve. The problem is easily solved, as we have seen, with the aid of logarithms. After an adaptation that a good artist should be capable of making, the musician could play a piece in such a scale; string instruments and some wind instruments could be tuned in accordance with this scale. We do not underestimate the difficulty, but men have surmounted worse ones. Such an effort would at least be undeniably original in character, and perhaps one of the scales which could be obtained in this manner is that of the nightingale, or of birds in general.

Let us note by the way that musical notation is nothing but a graph with two variables, since it indicates the length of a note and its pitch.

Conversely, a graph drawn on a musical staff can be played. Let us draw a curve on a musical staff: The points where this curve inter-sects the lines of the staff shall define a note whose length can be indicated by the horizontal distance which separates two neighboring points of intersection. Mathematical considerations might well be an aid in plotting such curves.

To sum up, 2500 years ago music was identified with the mathe-matics of that period: The Pythagorean school created them both on the same principles. Then there was a slackening in the relations be-tween music and mathematics, the tempered scale of 250 years ago being the last important result of their collaboration. There is no reason to conclude that we have seen the last of this collaboration, in fact there are unquestionable possibilities which future musicians will perhaps put to use.

Book Five

MATHEMATICS AND TECHNOLOGY

Mathematics is part of the vast enterprise of "domination of nature"[1] by mankind. No fact can shed as clear a light on the essence of abstraction, as well as on one of the reasons for its existence. Abstraction, which as its etymology reminds us, is extracted from the concrete by judicious eliminations, justifies itself and confirms its validity by the power it confers upon man.

The usefulness of mathematics is as universally recognized a fact as is its truth. Thus we have not believed it necessary to dwell on this point at the expense of less important, but also less investigated areas of our inquiry. We are nevertheless sorry that we have not been able to demonstrate in this volume the direct and protean penetration of mathematics into our daily life. Our thoughts and actions are so thoroughly infused with mathematics that we are scarcely aware of it. Paul Montel writes:

> Animals, too, obey it, and their instinct, sharpened through the slow working of heredity, has led them to discover mathematical laws, which seem in some obscure fashion to inhabit the very structure of their consciousness.

To this sound observation one could add that man himself is unaware that he remains an animal whose activity is based to a great extent on intuitive mathematics. The simplest act, reaching for an object and manipulating it, and the most elementary conclusions we draw concerning collections of things, introduce a geometry and an arithmetic that are known, so to speak, to our bodies, their

[1] The expression is Descartes's.

organs, fibers and even their cells. At a higher level, agriculture, the crafts and commerce reintroduce the same disciplines under the more elaborate categories of measuring, drafting and counting. It is at this technical level that mathematics enters the sphere of the conscious and is impelled towards science.

But the extraordinary amplification of mathematics brought on by its role in the exact sciences ("exact" often signifying mathematized) is the main reason for its participation in the most advanced—and sometimes the most troublesome—aspects of human activity. The following two articles open horizons upon a question about which perhaps less is known than appears to be the case.

At the same time a theorist and a technical expert, a specialist in problems of mechanics and thermodynamics, Maurice Roy is the model of a great engineer. He is one of our most knowledgeable experts in the field of jet propulsion, which to a certain degree dominates present political activity. His article, concise but weighty, reveals the difficulties that engineers encounter when mathematical formulations are not adequate for their purposes. At the same time his paper allows us a glimpse of some of the elements of a solution, products of the author's lengthy personal experiences.

Michel Luntz, a distinguished scientist in the field of hydrodynamics, discovered through experiment, a new type of turbulence within fluids thrust into an electric field, and has given a theoretical interpretation of this phenomenon. Although he was interned in a concentration camp at Vernet, he pursued his studies on the iteration of functions, and succeeded in getting an elegant theorem on this difficult subject into the hands of Jacques Hadamard, who was then in the United States. His article comprises a general examination, illustrated with numerous powerful examples, of the relationships between mathematics and industry.

F. LL.

45

MATHEMATICS AND THE ENGINEER

by Maurice Roy

PROFESSOR, ECOLE POLYTECHNIQUE,
CORRESPONDING MEMBER OF THE INSTITUTE

IN the art of engineering no problem can be solved without resort to mathematics, a fact which justifies the rather large place given it in the training of engineers.

Moreover, in the 19th century, there were a considerable number of engineers, accomplished mathematicians, who personally contributed to the progress of geometry or analysis—doubtless pressured by some necessity.

But this state of affairs, especially since the beginning of this century, has developed along two different lines. On the one hand, the field of mathematics has expanded considerably, and it is no longer possible except in very unusual cases for somebody effectively devoting himself to progress in the art of engineering to excel in mathematics. On the other hand, engineering itself has in all its branches become more diversified and complex, at the same time that it must resort to ever more refined theories.

Thus, in a very general way, the expansion and complexity of the knowledge needed by the modern engineer no longer allows him the possibility of being sufficiently expert in the varied advances of mathematics, precisely when he has greater and greater need of its most advanced procedures to solve ever more difficult problems.

To be sure this is a hackneyed statement and nobody seriously denies that intensive specialization is increasingly necessary at the professional level. But the observation has some bearing on the subject we are dealing with here.

Mathematics is not only an excellent and indispensable intellectual exercise for the engineer during his training period, giving him a taste for rigor in his thinking, but it should also permit him to understand and follow the progress of the physical sciences with their constantly increasing mathematical content. Finally, it must also prepare the engineer to extend the theories of these sciences far enough for applications that interest him.

This last objective in particular seems to be almost completely impossible these days. Indeed, however much we may augment the engineer's mathematical training, it can no longer be universal enough nor intensive enough, particularly after the passing of time has made inroads on this knowledge, to empower him to learn a difficult theory from scratch and completely solve a concrete application of it. The cooperation of the mathematician and engineer, combining the unique competencies of the two, therefore becomes more and more indispensable.

Before any further words on the subject we must emphasize the very special character of the engineer's use of mathematics. Assuming that he has correctly formulated a problem, his most important concern is to obtain numerical solutions. Usually these are obtainable only by methods of approximation.

In general the mathematical training given to future engineers does not allow sufficient time for discussion of these methods. If the goal is to fix ideas in students' minds, should the partial differential equations that automatically come up in the mechanics of continuous and deformable media be gone into? We teach, rather, the principles of the solution of such equations in certain very special cases, for which, besides, it is often necessary to resort to special functions. In the case of these latter, it is mainly their properties which are analyzed and the student is not shown the way to practical applications. In fact real technical problems do not generally fall into a category possessing a classical solution, and the student is given no guide for solving the problem in question in a correct, approximate fashion.

For lack of well-grounded training in approximation methods one too often sees engineers arbitrarily simplifying equations in order to make them solvable or inventing procedures that give an approximate solution but fail to have any correct foundation. Thus arise the unsatisfactory theories whose basic inexactitude causes so many engineers to discredit the effort to treat a technical problem scientifically, no matter how meritorious this is in itself.

It would be very useful to devote part of the mathematical training of an engineer to this very necessary art of solving the principal types of equations by correct approximations. But this is a particularly difficult kind of teaching and must be accompanied by numerous exercises, contrary to the usual pedagogical methods of engineering schools.

As a matter of fact, pure mathematicians are interested in methods of approximation only if they need a practical solution of an equation, and this necessity occurs, in general, only when physicists or engineers with whom they are associated express a need in a particular situation.

One of the useful results of cooperation between mathematicians and engineers, the absolute necessity of which we mentioned above, would be to lead mathematical experts themselves to elaborate and perfect methods of approximate solutions; at present they still seem to be too ignorant of the utility of such solutions. Is it not this, by the way, which so unjustly causes this area of study to be discredited?

The desired cooperation is in truth very difficult to bring about. All the more reason to work for it indefatigably.

A major difficulty is doubtless the lack of contact between individuals in the areas concerned, a useful remedy for which would be the creation of centers of higher, specialized learning; these would bring academic people and engineers together in teaching and research, in close conjunction with laboratories whose activities should infuse some life into teaching itself. This kind of desirable innovation was undertaken in 1947 with the establishment of the Center for Advanced Studies in Mechanics under the aegis of the University of Paris.

Another difficulty arises from the fact that each problem, by its very nature, relies on the competency of only a small number of mathematical specialists; the engineer involved in the problem at hand either cannot learn their names or cannot hope to reach them because the lack of personal relations does not permit it.

Many engineers finding themselves in this situation dream in vain of an ideal intermediary who could interest the appropriate specialist in the mathematical difficulties they have run into.

Is it perhaps showing too much confidence in the inclination both groups have to be of service to hope for some organizing effort that will facilitate the desired personal relations?

Recent computer advances give new urgency and new value to this organizing effort.[1] In fact we know that there are machines today

[1] On this subject see footnote 18, page 40, of Boll and Reinhart's article. (Note by F. LL.)

capable, for example, of calculating integral surfaces satisfying boundary conditions and initial conditions for very complex equations. But the cost of these machines makes their use possible only if we equip a center and organize it to meet multiple demands.

If one may be permitted to make a suggestion, where in our country would such a center be better located than right in our National Center for Scientific Research?

By entrusting it to the mathematicians, we would be assuring a solid basis from the outset for the cooperation so many engineers desire. At the same time the specialists, who do not at present have the time to solve the equations and therefore limit themselves to analyzing the conditions for the existence of a solution and at most certain properties of the latter, could advance their studies in an eminently useful direction and be themselves led to new discoveries.

These circumstances could bring about again relationships like those which have frequently occurred between physics and analysis, and the satisfying of the essential needs of the modern engineer would contribute to the progress of mathematics itself.

46

MATHEMATICS IN INDUSTRY

by Michel Luntz

OUR daily life is permeated by mathematical thinking, which reaches into every aspect of our surroundings. Sometimes trivial, sometimes unbelievably complex, it escapes our attention; we are too used to it to notice it.

You flip the switch and the light goes on. In a distant hydro-electric station turbines, whose tiniest element required dozens of hours of mathematical calculation, start up the dynamos that generate the current. It took years of calculation to design these dynamos. The generators send the electric current through transmission lines. It was a new task for the mathematician, and a particularly arduous one, to solve this seemingly simple problem of transporting energy. Beyond the application of Kirchoff's simple formulas and the difficult problem of the equilibrium of the network, it was also necessary to calculate the entire system of cable supports and the elastic tensions to which the latter are subject. They must not break nor sag too deeply under their own weight or through the action of wind or snow (the very simple problem of the catenary). And now the electric current is at your door. It flows through the filament of the lamp and brings it to incandescence. The light gleams. It is simple, but there again, without the aid of mathematics, what would happen is simply that all the electric wires in your house would heat up and two or three times a year the firemen would come to put out the fires produced. The filament of the lamp would barely redden, and if unfortunately it should begin to glimmer with more intensity, it would mean that your bulb was about to go out.

Let us take a more commonplace example: the house you occupy. Compare it with that of your ancestors (wherein mathematics already entered to a great degree, but at a level limited to elementary geometry and arithmetic). Your walls are not as thick; however, the house stands up as solidly if not better than the older houses. Yet the materials employed are less solid or at most equal to the old materials such as square-hewn stone. Mathematics has come along and made man's work simpler and his life more comfortable.

When the bicycle was invented, its inventor unconsciously employed principles of rational mechanics that were rather simple but on which the greatest mathematicians of mankind have left their traces.

We are familiar with the example of savage tribes whose counting does not go beyond five. Numeration is the simplest and most fundamental operation of mathematics. It is difficult to build other operations without it. It is not by chance that the material and intellectual life of these tribes is at a rudimentary level.

One can live comfortably without music and literature; it is impossible to do so without mathematics. Mathematical thinking is not only a system of reasoning, an abbreviated language; it is even more an instrument for constructive thinking.

In the past, the use of mathematics was confined to arithmetic and elementary geometry, the only mathematical disciplines that were well assimilated. Among the mathematical formulas employed in the life of very ancient peoples, one should cite the calculation of surfaces and volumes, the laws of the lever and Archimedes's principle. The development of navigation became possible only with the use of trigonometry. Then came the times of Newton and Huygens (a little-known specialty of the latter, was the design of barouches and sedan-chairs). The invention of logarithms by Napier, the laws of motion by Newton, the differential calculus by Leibniz, this is the true revolution that brought humanity into the century of the machine, which profited in addition from the mathematical theories of Carnot and Ampère.

But while first the physicists, and then the engineers, immersed themselves in mathematical principles and built steam engines, locomotives and steamships, the mathematicians in the advance guard of human progress, Gauss, Cauchy and Riemann, made a breakthrough to a new stage in knowledge. Based on this new advance, Maxwell's discovery of the equations of electromagnetic propagation led to today's electrical and electronics industry. Modern conformal mapping and modern aerodynamics are based on Cauchy's theory of analytic functions, and no airplane would be flying if the theory of analytic functions were still unknown.

As one of the most curious instances of the success of pure mathematics, we should cite the construction of "laminar" airfoils (whose resistance to forward motion is several times less than that of the usual airfoil); these are based on recent very complicated, purely mathematical work done by the English professor Goldstein.

Levi-Civita's tensor calculus led to the theory of relativity. The theory of relativity leads to nuclear physics, and this leads to the laboratory at Los Alamos and to the atomic bomb. Is it not strange that the cemeteries at Hiroshima should owe their existence to those most peaceable idealists and mathematicians, Levi-Civita and Einstein? In the meantime, that same atomic energy cures cancer, transforms coal into diamonds, drills tunnels under the sea and carries us into interplanetary space. Will the theory of spinors and the theories of Hilbert spaces lead us still further into the domain of technical applications? The unforseeable future will answer these questions. But there is every reason to believe that the answer will be positive.

If in the majority of cases mathematics preceded physics and the latter preceded industrial applications, we also are familiar with some examples of a flow in the opposite direction, where problems posed by industry induce, at the end of the cycle, improvements in mathematics.

Today there are a great number of automatic devices that solve very complex mathematical problems. These devices exist because industry benefits from them, but numerous mathematical problems incidentally find an *approximate* solution. The most striking example of such a device is the electrolytic tank, perfected in France by Perès and Malavard, which solves the problem of harmonic functions with the values given on the contour.

Other very complicated devices solve systems of linear differential equations up to the sixth order. Still others automatically calculate artillery range-tables. It is almost unnecessary to mention the slide rule and the classic calculating machine whose field of action is limited to the fundamental operations. In America there are also slide rules which perform operations on complex variables. The planimeter, which is widely employed, carries out the integration of a function of one variable represented graphically. Another device widely used in certain industries is the harmonic analyzer, which develops a given function in a Fourier series and automatically calculates the successive terms.

In a general way, one can construct a calculating machine to solve any problem whatsoever that corresponds to a well-verified physical law. For example, if we wish to construct a calculating machine to give the inverse of the square, we could use Coulomb's or Newton's law of attraction and repulsion. If we wish to integrate a linear differential equation of the second order we could construct a calculating device involving an electromotive force, a resistance, a self-induction coil, a capacitor and an instrument to take the appropriate readings, etc.

Thus industry can, if the need arises, solve a great many of its mathematical problems by itself, with mathematical robots. To the extent that the level of the mathematics entering into the formulation of physical laws, and thus into industry, rises, the level of mathematical problems whose solution becomes possible through physical methods rises at the same time.

This development in technique which is stimulated by its own advances—a sort of chain reaction of progress—has great bearing on the future, providing, however, that economic and social conditions do not check it, as is often the case.

But to solve a mathematical problem is often a very simple thing when this problem has been correctly posed. We believe that this is where industry's greatest weakness resides in so far as the rational employment of mathematical thinking on new problems is concerned. By its very structure, which is generally oriented toward an immediate profit, industry employs only the staff strictly necessary to perform its function of realizing these immediate profits.

The manufacturer prefers to replace a qualified worker with an unskilled laborer; he prefers a foreman to an engineer, an engineer to a physicist, a physicist to a mathematician. And while the engineer is paid better than the foreman, the physicist is paid less than an engineer, and the mathematician is the physicist's poor relation. Thus the economic structure of society is inherently unfavorable to pure research, that is, research whose results are not immediately translated into material profit, but which produces deferred results that determine the whole of human progress. A rational determination of the proportion of the staff to be devoted to pure research, and thus to mathematics, to applied research, and finally to translating applied research into an existing reality, should take precedence over the contingencies of immediate profits. Moreover, a new category of employee should be created, namely, that of industrial mathematician.

The industrial mathematician should possess extensive mathematical knowledge in all branches of the subject which have already had applications, and also those for which one can foresee applications in the near future. He should be able to refer problems arising from applications to the already known chapters of mathematics; on the other hand, he is not barred from inventing new theorems or new mathematical theories himself. Such men, by trying to reduce the constant lag that exists between pure research and daily life, would play a very important role in the economic life of society. In many areas of activity this lag can be measured in decades; in certain areas, it may even be centuries. We could even speculate about creating special institutions designed to "prospect" for these persistent anachronisms in the various branches of industrial activity that escape the observation of the trained professional because he has pushed his specialization too far.

Book Six

MATHEMATICS AND CIVILIZATION

Through education, philosophy, the sciences, the arts and techniques, mathematics makes a considerable contribution to the shaping of civilization as a whole in its economic, social, political and religious aspects. This osmosis is not a one-way process. Mathematics is affected as much as it affects. Sometimes visibly and ruthlessly, more often indirectly though decisively, the forms and development of economic and social structures condition the superstructure, at whose summit the ensign of mathematics proudly flies. This is still a new subject whose mastery would require real mathematical competence coupled with a broad cultural background. One must above all be suspicious of overly simplistic schemes. The processes of abstraction manifest a sort of inertia and are like the phenomena of hysteresis (in the sense in which these two words are employed in mechanics and electricity). Certain superstructures can flourish in opposition to the environment surrounding them: Rates of infiltration and propagation are not infinite, so that simultaneous causes can produce successive results and conversely.

The great movement known as the Renaissance had its beginnings, in the realm of art and manners, in 15th-century Italy. In our judgment its equivalent on the mathematical level is the 17th-century discovery of differential and integral calculus in the Northern lands. Thus it took two centuries to go from the social to the mathematical, and from the shores of the Mediterranean to the North Sea.

Like flowers plucked from the nourishing soil, branches of human activity or knowledge live for a time with a life of their own, at varience with the entire environment enveloping them. They continue to develop by virtue of an acquired velocity which is communicated to them at the start through fortuitous and

210 EDITOR'S COMMENTARY

accidental causes, such as the influence of a man of genius. Of course in the end they become anemic and disappear if they do not receive nourishment and stimulus from the society in which they sink their roots. This is probably the reason why the Hindus men of the Middle Ages, both Byzantine and Occidental, each so preoccupied with the infinite in their theology, showed themselves incapable of rising above a heady but confused point of view to a scientific stage. The same process explains why the Greeks, incomparable logicians but too often indifferent to the lessons of experience, were able to extract from the arguments of Zeno of Elea nothing but philosophical paradoxes without any scientific or technical implications. A genius such as Archimedes, who in many respects is closer to the mentality of 20th-century man than to that of his contemporaries, was indeed able to clearly anticipate the infinitesimal calculus; nevertheless, he could not get his contemporaries to adopt it. Which shows how indispensable it is to secure general agreement on an idea in order to assure its triumph. How could this be established without a favorable economic climate?

A second proof of this argument is furnished by the fact that, subsequent to these abortive efforts, there did arise a powerful body of theoretical knowledge in the field of infinitesimal analysis. It required the change from a feudal economy to a capitalist economy to ensure this triumph. No denigration of the striking genius of a Newton or a Leibniz is implied in pointing out that not only did each of them find the principles of the new method independently (each in a different manner, according to his personal temperament), but even more, that they were preceded by a multitude of contemporaries, Cavalieri, Barrow, Fermat and others, who came within a step of the final synthesis. The time was ripe. If Newton and Leibniz had not been born, society would have ended up creating the infinitesimal calculus without them, only a bit later.

One could doubtless discover similar explications regarding the creation and development of each of the great branches of mathematics: geometry in ancient Greece, algebra and the theory of algebraic equations at the beginning of the Renaissance, the calculus of probabilities at the dawn of modern times, the theory of groups, the theory of sets, the algorithms of vectors, matrices, tensors, operators and symbolic logic in our times. We hope that competent historians of mathematics will show an interest in these absorbing questions.

Jacques Chapelon, who is very well informed on mathematics in the English-speaking world, has given us works on ternary quadratic forms (in the theory of numbers) and on mathematical statistics. Based on a wide knowledge of mathematics and on a just appreciation of its relations to social developments, his article picks up the main threads in the development of mathematics, already sketched out for us by Germain. But this time he systematically thrusts mathematics into its human setting and thus reaches back to the causes of this dazzling *pyrotechnical display, the first flares of which—to our knowledge—shot up simultaneously with the passage from tribal organization to great empire. This text is a model of its type; it constitutes a valuable, first contribution to what we hope will be a long series of investigations.*

F. LL.

47

MATHEMATICS AND SOCIAL CHANGE

by Jacques Chapelon

PROFESSOR AT THE ECOLE POLYTECHNIQUE

In studying a paper by one of his Japanese colleagues, a French mathematician can be totally ignorant about Japan and the author's personality. If the work is written in a language accessible to the reader and conforming to ordinarily accepted notation and conventions, a mathematical bond will be found, and the work of the Japanese mathematician will become a source of reflection for the Frenchman, who will follow up its implications. The very style of mathematical works tends to be bereft of any social or human aspect. Author's names are often given without any clue as to whether they are dead or still living. At most one now and then says that a proposition is "beautiful" or that a proof is "elegant." Thus one sometimes hears it asserted that the activity of mathematicians is independent of the society in which they live, but that mathematics, because of its repercussions on technology, does have an effect on social change. Therefore, goes the argument, the development of our civilization takes place on a mathematical framework resulting from a sort of revelation that transcends sensible reality. It is recognized, to be sure, that mathematicians continue the investigations of their forerunners and are therefore partly dependent upon them. But, it is said, they create arbitrary entities; they make definitions they consider appropriate; and from that point on everything reduces to a mechanical, syllogistic undertaking leading to that magnificent chain of propositions that a mathematical theory brings forth. According to this line of thought, mathematicians constitute a world sport; they tend to behave like pure spirits, communicating to other pure spirits their meditations on the eternal verities. And one is thus led to view mathematics as no more than an indefinite development of logical speculation constructed on a substratum of arbitrary conventions; like Allah the all-powerful, who creates the entirety of the world each time a quantum of time drops into eternity, the mathematician-creator arbitrarily molds through his work the indeterminate future of his science. This picture is reinforced even more when it is asserted that the mathematician's aloofness from any concrete reality does not prevent him from producing a valuable piece of work, and what is rather surprising, one rich in practical applications, even though he professed to be uninterested in such results. Consequently, it appears, mathematicians have an influence on the state of their society, but, inversely, the source of their subject's progress is the genius of the great mathematicians. And thus we come back again to the view that the driving force in social change comes from men of great learning and men of genius.

These views result from a superficial analysis of reality. We accompany the mathematician in his apparent isolation. Accepting this isolation as a primary and irreducible given, we then establish that mathematical production occurs. This very naturally leads us to the idealist conclusions that Henri Poincaré summed up so well in his celebrated statement: "Thought is nothing but a flash of lightning in the middle of a long night. But this flash is everything." Reality is different. One cannot pass a valid judgment on the development of mathematics if one arbitrarily isolates this development from its context and its background. The process of such a development is much too complex and much too intertwined with the general evolution of humanity; it becomes so completely distorted, to the point of unintelligibility, by being thus isolated. The human and social character of mathematics cannot be disregarded, for mathematicians and the societies in which they develop form an inseparable whole. It is, in fact, by reintegrating mathematics with social development that one can succeed in understanding how, arising from the technical needs of society, mathematics has little by little gained prodigious dimension and sovereign pre-eminence, and finally how, through one of those frequent turnabouts in history, it has become one of the fundamental ideological foundations of civilization in contemporary society. Thus we must look to history to gain some understanding of the interactions between social and mathematical development. We shall limit ourselves to a brief study of some examples of these interactions, for in this short article we cannot hope to present a history of mathematics.

Mathematics was born when the material needs of life required its existence and when the technology of some society had attained a certain level. At the beginning it had only a prescientific, empirical character. It next reached the experimental level, that of a true physical science, a physics of number and form. Let us take the cardinal number. If you go beyond the first few numbers, it is scarcely grasped by direct sensory means. Languages of primitive peoples sometimes have no word to designate a quantity larger than four, however rich or redundant they may be as far as concrete details are concerned. After four they say only, "There are many." The notion of number arises from the technical need to attain the cardinal number. The first mathematician was perhaps an ingenious herdsman who wished to count the animals in his herd and conceived a technique of enumeration or correspondence which, basically, amounts to getting at the cardinal number through the intermediation of the ordinal number.

The most rudimentary of agricultural economies needs numerical information concerning the seasons. This implies the solution of problems connected with the construction of a calendar. We know how widely problems of calendar-reckoning, and consequently problems of astronomy, were studied in the most varied of primitive civilizations. Next, the decoration of the human body, tools and instruments, the potter's art and the architectural preoccupations which arose when man began to build, implied certain geometric considerations; these often remained at a purely empirical stage, but at other times attained a higher level.

Mercantile societies have an immense historical importance. Our capitalist society is itself the great-grandchild of the mercantile economy which was grafted onto the feudal economy in the Middle Ages. Now, a mercantile economy in its beginning stages, particularly if it is a maritime economy, poses a multitude of technical problems which thrust themselves insistently upon men's thoughts. Such an economy needs a bookkeeping system, rules for dividing inheritances, the art of navigation, means of transportation, in short, a whole body of techniques requiring the use of rudiments of the theories of arithmetic, geometry, astronomy, mechanics, etc. This development takes place in every mercantile civilization. It is how arithmetical and geometrical techniques arose in Egypt and Babylonia, and pre-algebra in India. We can also cite the very special needs faced by the Egyptians in re-establishing the land boundaries periodically erased by the mud of the Nile floods. At its dawn mathematics is thus closely dependent on the technical level of the society.

In their later development these mathematical techniques became rather complex and no longer accessible except to specialists. These specialists, gathered together by the ruling classes, became a caste in the state's apparatus. Mathematics thereupon took on an arcane character. Mathematical activity became the prerogative of certain initiates. The men who were in possession of "hidden things"—the expression used by the scribe of the Rhind papyrus—enjoyed a monopoly of knowledge and this gave them vast power. (We are just beginning to break this monopoly and to assure a democratic dissemination of the sciences.) Far indeed from the mathematician isolated in his ivory tower, speculating freely upon subjects of his own choice! But this arcane quality of mathematics led to the development of a mysticism of number and form, and a like mysticism appeared in the Mediterranean basin as well as in China or among the Negroes of the Congo.[1] The importance of this mysticism should not be underestimated. We know the influence of Pythagorean mysticism upon the development of Greek science and philosophy, and its propagation from generation to generation including our own.[2]

In Greece the introduction of Egyptian papyrus was the prelude to the birth of mathematical thought. This is an instance of social conditioning of a technological and economic nature. Greek society was essentially a commercial and slave society topped by an aristocratic democracy of citizens. A society based on the work of easily obtained slaves, in which increasing the yield by technical improvements was scarcely a matter of importance, quite naturally brought about the withdrawal of the ruling elite from concrete reality. This social structure imprinted a very original character upon Greek mathematics, namely, a disdain for practical applications. And one is not surprised to see that a member of the Greek ruling class was possessed of leisure, had a taste for travel and was inspired with a love for intellectual speculation. The Greeks' alienation from material reality led them to deny the value of this reality. It is the world of ideal essences which the sage must strive to attain; the reality of the senses belonged only to the world of appearances and was no more than a debased reality,

---

[1] Pelseneer, *Esquisse du progrès de la pensée mathématique* [Sketch of the Progress of Mathematical Thought], Paris, 1935, pp. 28–30.

[2] For example, Kronecker's saying: "The good Lord created the integer; all the rest is but the invention of men."

a tottering support of the reality of the essences. The structure of Greek society is the material basis for the Greek taste for abstraction and reasoning. It is correct to say that it was also the foundation of their rationalism, their confidence in the power of pure reason to attain truth and their admirable technique of proof; this last point was to be of fundamental importance for the subsequent development of mathematics. Counting upon reason alone to further the progress of mathematics, the Greeks knew how to employ it with irresistible art and power. We shall have to wait for Roger Bacon to suspect that progress in other sciences, at least in their initial stages, will result from a less rigorous rationalism, and that truth can be attained, at least approximately, by more efficacious and more certain procedures than the geometric reasoning of the Greeks. In turn, the unparalleled successes of their geometers encouraged the Greeks to shun sensible reality all the more. Repudiating the profound conclusions of Heraclitus, the philosopher of flux, they reached the point of wanting to grasp the reality of eternal and fixed essences. Then again, according to the Pythagoreans, truth, beauty and the good must be sought in unity, in the finite and in repose. The conjunction of these tendencies explains the static character and immobility of the speculations engaged in by the Greek geometers. In one of the arguments of Zeno of Elea, an arrow is at rest at every instant and its motion is denied. Even for Euclid, in the Alexandrian period, the circle is not considered the locus of a moving point, as it readily is for us; some other definition not requiring the introduction of motion is given. Linked to this requirement of immobility is also the surprising claim of Greek geometers that the only acceptable construction is one requiring no other instrument than the compass and straightedge. We have here a Pythagorean prejudice for the preeminence of the straight line and circle, superposed upon an appreciation of the fact that in use the straightedge and compass are solid, undeformable bodies. This anti-dialectical character of Greek geometry, associated with Pythagorean ideas preaching the preeminence of the integer, the abomination of the irrational number and the exclusion of mathematical infinity, prevented a development of mathematical analysis. However, the Greeks did possess rudimentary elements of analysis, which they got from the Babylonians. They themselves had conceived the idea of infinitesimal procedures but had never used them without some repugnance, and only as an imperfect research tool rather than as a regular and valid method of proof. To the credit of Greek mathematics are the astonishing beauty of its geometrical contributions, its demands for rigor, to which our own mathematicians submit as being in the best interest of the solidity of their research, and its confidence in the limitless power of human reason, even though modern rationalism has assumed a narrower character than that of the Greeks. The weak points in their contribution are their rigidity, their divorce from practicality, their inability to rid themselves of the limited concepts of Pythagoreanism and their failure to appreciate the scientific value of approximate truths.

In the different social environment of the Alexandrian Greeks, which was perhaps more mercantile and definitely more oriented towards mechanical studies, the character of mathematics develops appreciably; the Alexandrian mathematicians, however, never abandoned the prejudices and traditions of the Platonic period. Archimedes was one of the greatest mathematicians of all time. Trained in the Alexandrian school, he was not a gentleman of leisure but a great engineer as much as a great mathematician. Responsive to the social needs of his city and a distant precursor of Monge and Carnot, he found the creative elements of his work in social realities, and this gives his work an almost modern air, for he employs infinitesimal procedures, which makes him the originator of integral calculus. But Archimedes did not use infinitesimal methods lightheartedly; he too considered them only a makeshift. Later the great Diophantus, the man who anticipated algebra, was to write a celebrated work which perhaps did not remain unknown to the Hindus. But neither the great Archimedes nor Diophantus succeeded in creating a practical symbolic representation of numbers. Yet such a representation was necessary for the development of modern mathematical analysis, although this necessity is historical rather than theoretical in nature.

In India we find a mercantile civilization which quite naturally nurtured the development of commercial arithmetic and geometrical studies. The Hindus shared neither the Greeks' logical preoccupations nor their concern with rigor. They had no prejudices against irrational numbers; and they succeeded in establishing a better pre-algebraic mathematical system than had the Greeks. But their great contribution to mathematics will remain the very humble discovery of an altogether elementary symbol: In the first centuries of the Christian era an unknown Hindu conceived the *zero of position*. This discovery was foreshadowed by the Babylonians and even by the Mayans, but it was definitively formulated by the Hindus. This gave the world the mathematical technique which was to be a powerful aid in the development of mathematical analysis. The Greeks had indeed conceived a sort of analytic geometry, for example, but nothing came of it comparable to that which Cartesian analytic geometry was to produce. Their numerical analysis and their diagrammatic, symbolic representation of numbers was not satisfactory, and this was a general defect in all of classical antiquity. Archimedes in his *Sand-Reckoner* had striven to construct a very large number, but he had not manufactured a system of symbols permitting addition other than with the ancient abacus, and this instrument was practically the only method used by the Greeks to carry out numerical calculations. It was obviously inadequate for making a calculation with a given arbitrary approximation, which is the ultimate goal in the mathematical solution of a problem. A thousand years were to pass before a new technique was commonly adopted in the Western world, and it is this technique which, thanks to a favorable environment, was to transform Europe into the "continent of calculators" that Burke was to deplore much later.

The Hindu system of numeration first traveled to Islam, another mercantile civilization. Introduced by the Muslims into Europe, Hindu numeration was favorably received by the commercial world. In 1228, Leonardo Fibonacci of Pisa wrote the first book on financial arithmetic. In the 13th century Italian merchants were generally using the so-called Arabic numbers, this despite the interdictions of religious authorities. Later schools for arithmetic were created in Germany. Immediately after the invention of printing, books on commercial arithmetic were being printed. Our system of written numeration is thus a relatively recent acquisition although it forms part of the scientific equipment of our very young children. It is clear that this system was adopted under the peremptory pressure of social change. What was the origin of the irresistible impetus inspiring this change? The feudal economy was a closed economy wherein commercial exchanges were reduced to a minimum, and hence where the role of money was insignificant. The Crusades radically modified this somnolence. The lords who had tasted the sumptuous Oriental life came back with new needs. They had to procure money in order to buy objects of luxury, and as a consequence payment in money was gradually substituted for payment in kind. Markets and fairs arose. This posed numerous technical problems notably those connected with operations of exchange and credit, as well as with laying out commercial land routes. There was a constant impetus from the Italian cities on the Mediterranean coast, and from the Hanseatic cities, whose trade was essentially maritime, to improve all the techniques relating to navigation. The desire to find new maritime routes to reach the spice countries accelerated technical developments by posing the great problems of transoceanic navigation. These intense technical needs obliged men to concentrate their thoughts on seeking solutions to these problems. Here is where we can find the source of the potential for the development of modern science in general and of mathematics in particular. Ever since that distant era, this potential has renewed itself with each generation, for social development constantly poses new technical problems which always stimulate new scientific research. At the beginning of the 15th century a Portuguese prince, Henry the Navigator, created a school of navigation. Astronomical tables constructed by Jewish scholars using the Ptolemaic system rendered transoceanic navigation possible. The problem of position at sea presented itself. Only the development of a body of astronomical knowledge, including the creation of celestial mechanics, could satisfactorily solve that problem. Celestial mechanics led immediately to the creation of dynamics. There had been speculation from the end of the 15th century about using the moon to solve the problem of longitude, and a theory of the moon is one of the most difficult problems of celestial mechanics. Finally, other techniques stimulated fundamental mathematical advances, notably techniques of military arts, whose ballistic problems could be fully solved only by a new mechanics, Galilean mechanics. These are the most important of the elements that stimulated modern mathematicians. Here again, we see that the social change that began in the 15th century is what made the development of modern mechanics necessary. For Greek statics was incapable of solving the problems posed, and the foundation of modern mathematical analysis was a prerequisite for the needed dynamics. Thus one can see arising in a distant future Descartes and Galileo, Leibniz and Newton, and even Lagrange and Laplace.

This rapid sketch shows that the beginnings of mathematics are like those of any science; the pressure of social needs gradually raises what was originally only a collection of empirical prescriptions to the level of scientific speculation. Thus the initial development of mathematics is conditioned by the productive forces of a society constantly in transition. The influence of these productive forces continues beyond the initial period and dominates the whole history of mathematics. Particularities of mathematical progress correspond to particularities of social development. Indeed, there is a close parallel between social progress and mathematical activity; the socially backward countries are those in which mathematical activity is nonexistent or almost so. However, once the initial impetus is imparted, the relations between the two developments become more complicated than a simple one-sided conditioning, for mathematical advances in their turn have an ever more powerful effect upon social development. The most advanced sciences and the most important industrial techniques tend to assume a more and more mathematical structure. They make use of existing mathematical results, but also pose new questions ever more urgently. They sometimes mark time if mathematicians are not prepared to provide immediate solutions. They constantly demand fresh progress. Even more, they tend to modify the traditional thought of mathematicians. Thus they orient mathematical thought towards analysis of the discontinuous; they inspire it to abandon the domain of the necessary in order to attempt to construct an analysis of the uncertain. This stimulation provided by the other sciences and techniques is thus incomparably invigorating. If it ceased to operate there would be reason to fear that mathematics would develop into a sterile scholasticism. One could say, on the other hand, that all mathematical progress implies increased knowledge of the real world.[3] The various parts of mathematics are closely connected with each other so that every new attainment has more or less distant repercussions upon the whole of mathematics, and thus influences technical progress and therefore social change. This is why it is fitting that we encourage mathematical research in all domains without allowing ourselves to be dominated by the viewpoint of immediate utilization. The speculations of the Greek geometers aided Kepler in his work, and the problems posed by the Chevalier de Méré have made their contribution, after many a twist and a turn, to the building of present-day physics.[4] Then, far from originating in the solitary and almost mystical contemplation of the great mathematicians, an era's mathematical

---

[3] "A discovery in analysis arises at the moment when it is needed to make possible each new advance in the study of phenomena of the real world." (Hermite, 1882)

[4] With respect to this subject, see Kahan's article, Part III, Book Three, pp. 104–117. (Note by F. LL.)

progress is closely associated with the activity of men of that era. These continual interactions between the activity of mathematicians and advances in science and technology have an oscillating character, one might say. But the oscillations are not regular. They are unstable, and far from damping, their intensity continually increases. They raise the level of both scientific production and technical power. In our present period, the rhythm of this process and the rapidity of this rise become more and more intense, accelerating social change and shaking up the status quo of the present social structure. Through this, mathematics is an important factor in formulating the society of the future. Finally, in conjunction with the other sciences and technology, mathematics constitutes the foundation of modern humanism, that scientific humanism which alone can give direction to the aspirations of modern man, of real, contemporary man; here again, mathematics is preparing the advent of the social structures of the future.

Mathematical activity thus results from the contributions of a dead past, but also and above all it is the result of social influences, aspirations, efforts and the general tendencies of living men, all made concrete through the forces of production, the nature of the relations among productive forces and the inherent contradictions in all social flux. Such are the elements which, through their impact upon the mental life of the mathematician, bring into being, usually unconsciously, his desires and passion for research, his dreams, fancies and impulses, which cause him to neglect certain problems and attack others; these elements account for the somewhat disordered character of mathematical creation and are a reflection of the turbulence of social development. Conversely, and through reciprocal action, mathematical advances influence social development by raising the technological and ideological level of society and by contributing to the development of productive forces. The creative power of mathematics is itself constantly undergoing change, sometimes increasing, sometimes decreasing, ever exhausting and enriching itself, and synthetizing, on an elevated, spiritual level, man's conscious or unconscious aspirations. Mathematics thus forms the solid base of scientific humanism, an ideology adequate to social development at its present stage. If one accepts these views, the development of mathematics is harmoniously integrated with that of society and completely loses its mystical character.

PART IV

Appendix

This book was ready for the printer, and I was on the point of turning it over to him when I learned that only a few days previously the manuscript of the article Léon Brunschvicg had written for this collection had been found among his unpublished posthumous papers.

His heirs, and in particular his daughter, Mme. Weill, to whom we are indebted for this fortunate discovery, did not know about this until early in 1948.

Thus the difficulties of every nature delaying the publication of this volume have given us the opportunity to recover a text whose existence nobody suspected and have had the fortunate effect of enabling us to include the article in the collection for which it had been promised. Inasmuch as we could not insert it in its proper place, between the Ullmo and Mouy articles, we were obliged to place Léon Brunschvicg's text in the appendix.

We wish to express our very sincere thanks to Mme. Weill-Brunschvicg, whose diligent searches have uncovered most valuable material for us, made more moving by the memory of the circumstances under which it was written.

F. LL.

48

DUAL ASPECTS OF THE
PHILOSOPHY OF MATHEMATICS

by Léon Brunschvicg

I

IF we pursue our investigations of philosophy and mathematics to their very foundations, we find ourselves driven again and again from conclusions about subtle and unexpected forms of reasoning, back to the deep roots and unchanging conditions of the process of thought. It is the task of history to uncover these reasons and conditions, while excluding preconceptions due to hindsight. The more a man imagines himself independent of history, the more, on the contrary, he makes himself its prisoner; indeed he exposes himself to a whole crop of chaotic, confused ideas and relationships, accumulated over the centuries, which have slipped into what seem to be impersonal and innocent usages of language. In this respect the experience of the Pythagoreans provides a very special and probably irreplaceable lesson, long ago as it may seem to us.

Our civilization is indebted to the Pythagoreans for the method which, at one and the same time, wins the deep-seated acquiescence of the intelligence and renders its universality incontestable. When, by representing numbers by points, they showed that the successive addition of the odd numbers furnished the law for the formation of squared numbers, they were extracting evidence of a perfect harmony—of a fundamental and radical equivalency—between what is conceived in the mind and what is obvious to one's vision. The centuries will add nothing to the brilliant and incorruptible richness of meaning that Pythagorean rationality conferred upon the word truth.*

To be able to state this without the risk of providing a pretext for equivocation or cheating, or without arousing any suspicion of mental reservations or excessive exaggeration is the sign by which *Homo sapiens* recognizes that he is definitely distinct from *homo faber*, and that he is henceforth the bearer of the value, truth, the value which man invokes in passing judgment on all other values.

After this, and despite this, it must be stated that Pythagoreanism nonetheless betrayed its own ideal. The triumph of reason should have been decisive; it was immediately compromised by a twofold weakness within itself, a double sin against the intellect. We cannot certify that even today the traces left by these faults are completely gone.

The Pythagoreans, who loved to proclaim themselves friends of wisdom, were unable to resist the temptation to make arbitrary generalizations and to go beyond the results they had achieved. Proud of having penetrated the internal structure of numbers, they desired the secret of the interior essence of everything to be revealed to them through the medium of numbers considered as qualitative entities or as vehicles of properties undetectable in the ordinary course of events. Thus 5, the sum of the first even number, 2, and the first odd number, 3 (unity remained outside the series), would be the number for marriage, because even is feminine and odd is masculine. No curbs will any longer restrain Pythagorean analogies or neo-Pythagorean ramblings. Ultimately, the triumphs of *Homo sapiens* with all their promises of rationality will only have served to put *homo credulus* with his primitive mentality back in the saddle, bowing to the confused fantasies of the mass imagination and responding to the fallacious authority of *hearsay*.

We might, however, have viewed this paradoxical spectacle merely as an accident peculiar to the Pythagorean school, had the productive triumphs of an exact science managed to split away from the dangerous speculations of metamathematics, and had the obstinate attachment to the latter not impeded the regular progress of the former. But this is not at all the way it was. What occurred was a conjecture of

---

* Here there is a break in the manuscript. Textual continuity has been restored by the insertion of a passage, framed in similar sentences, from the chapter "Reason" in Léon Brunschvicg's final work: *Héritages de mots, héritage d'idées* (Presses Universitaires de France, 1945), written at the same time as this article. (Note by F. LL.)

circumstances which over the centuries will continue to influence the course of mathematical thought; one circumstance concerns the nature of magnitude as related to number; the other is simply a play of words linked to an ill-omened peculiarity of the Hellenic language.

The first arises from the application of the Pythagorean theorem to the case of the isosceles right triangle. By adroit use of the method of *reductio ad absurdum* the Pythagoreans had succeeded in proving that the hypotenuse (or diagonal of the square) is incommensurable with a side of length 1: The hypothesis of commensurability would in fact lead to the conclusion that the quantity in question would be simultaneously even and odd. It seems that the existence of a magnitude which is absolutely and definitely present in space but which resists the ordinary combinations of pure arithmetic should have opened up a new and fertile field for mathematical exploration. Actually it is at this juncture that dogmatism transforms the original achievements of science by transferring them into a "beyond," thus shutting the door on an entirely different kind of leap, which would occur on a level of real knowledge and in accordance with the requirements of the rational method. To their marked disgrace, the philosophers of the school feel ashamed of what should have been the greatest title to glory in their mathematics. Their discovery of incommensurables inspires a religious terror in them by virtue of the fact that it shatters a magic charm and routs the harmony of a universe obedient to the authority of number. They receive a command from their avenging gods to deliver to the fury of the tempest the sacrilegious member who had the audacity to divulge the mystery of incommensurability. At this point the second circumstance enters upon the scene, with an effect that is peculiarly disproportionate to the cause. For the Greeks "logos" meant language but with the condition that it include the postulate, that the pre-established forms of the "logos" exhaust the resources of the intellect and permanently fix the frontiers of science; this postulate is all the more dangerous in that it remains implicit, for it will not allow any doubts as to its truth. Henceforth, the incommensurable, being something that one does not dare to speak of, will be considered irrational, and the confusion becomes inextricable; the specter of the irrational threatens to obscure the whole philosophy of science. We received proof of this in the debate which arose early in this century concerning the interpretation of thermodynamic principles, which was so brilliantly recorded in the work of Emile Meyerson.

From the ontological point of view, which was the position taken by early realism (and according to Meyerson, scientists, through a kind of animal instinct, persist in their attachment to this almost in spite of themselves), the conservation of energy would mean the persistence of something; it would thus be an expression of the mind's need to establish a standard of pure identity, although it would have to be admitted that this was a completely unreasonable need, since it would result in leaving out of consideration just that motile reality which it was its task to explain. In any case, reality rebelled by giving rise to a contrary principle: the dissipation of energy, or *Carnot's principle*; to put it better, the increase in entropy throws into relief the inescapable role of irreversible time. *Identity* and *reality* will therefore be in direct contradiction, at least as long as critical reflection does not enter and dispel even the appearance of a paradox. In fact, as soon as energy is thought of strictly in accordance with the intrinsic truth which gives it the right to be considered in science, it reduces to its mathematical expression, just as does entropy. By virtue of this the principle of conservation finds a natural complement in Carnot's principle, which, since Boltzmann, itself appears to be deducible as a special application of the calculus of probabilities. Finally, there remains the question of deciding whether this calculus itself still belongs to the domain of reason. Throughout its history, two types of thinkers have taken issue with each other over this question. There are those, from Pascal to Cournot, who ask mathematicians to combine finesse with a geometrical mind; and there are those, like Roberval or Auguste Comte, who deny them that right.

In our opinion no other example is more suggestive, and probably none is more decisive. It shows fully how the preconceptions of an overly abstract and narrow definition transforms reason into a machine for fabricating the irrational. But then one also wonders why reason should have consented to remain the dupe of terminology about which it had not been consulted and which it was obliged to use in order to give expression to itself and its activity.

Thus, behind the factitious opposition between the rational and the irrational, we discover a dual conception of reason and corresponding to it, two different ideas of mathematical philosophy. The first is a basic dogmatism wherein the philosopher relies upon an affirmation of transcendence which no one can be so audacious as to doubt or so scrupulous as to verify. Before venturing upon the scientific problem proper, he supposes that he already possesses the framework within which the actual research is expected to take place; he assumes the right to fix the character and the import of the results which research may lay claim to, and the right to furnish definitive answers to the questions which technical and experimental procedures of investigation leave up in the air. It will therefore be worthwhile for the philosopher to commit himself to a modest and serious pursuit of the steps taken by thought in its process of discovering problems and proving theorems. And the more original, bold and unforeseen these steps turn out to be, the more valuable will he deem them for his objective, which is to know the precise and true nature of the mind; for the mind is not bestowed fully developed; it fulfills its role only when it continually renews its effort to wrest itself from the products in which it would naturally tend to become crystallized and materialized. Hence the effort should be pursued without allowing the work to be impeded by this or that particular formulation of the principles which make his activity productive. It was with this view that Laplace wrote to Lacroix in 1792—toward the end of a century which had multiplied the theories destined to form the basis of infinitesimal calculus:

> The reconciliation of methods which you are taking into account will serve to mutually clarify them, and what they have in common usually comprises their true metaphysics; that is why this metaphysics is almost always the last thing to be discovered.

As long as number was in itself an object, the philosophy of arithmetic inevitably oscillated between primacy of the *cardinal* number and that of the *ordinal*, just as the philosophy of logic was incapable of settling the dispute between intension and extension. Here and there the confusions vanished when the static realism of concept yielded to the dynamic idealism of judgment. Number comes into being through an understanding of the operation by which a new and progressive image of *collection* is made to correspond to each successive act of *seriation*. But this had not been clearly and definitely recognized until after Georg Cantor provided the insights of his theory of sets. Even if it was still maintained that mathematics is founded on pure reason, independent of all recourse to experience, yet it was necessary to recognize that the priority of its content cannot form the subject of an a priori scheme, a transcendental divination. Certainly this is a paradox; hence it is all the more remarkable to note that those who stated it were in advance of the *law of awareness* which Edouard

Claparède was to formulate in 1918 and which he properly regarded as the most useful achievement of contemporary psychology.

> The earlier an individual's conduct implies a certain relationship and the longer he has made automatic use of the relationship, the later will he become aware of the relationship.

To push the exploration of the sources of automatism as far as possible, and by means of this to restore to thought the autonomy, thanks to which it will be able to soar beyond the frontiers that at first seemed to enclose it, such is therefore the task which mathematical philosophy undertakes, and which justifies it in so far as it is a discipline possessing its own conditions of existence.

II

If we have succeeded in describing the change in attitude that recurred—from dogmatic extrapolation to the analysis of critical thought—we shall find it possible to point out the phenomena of interferences that it explains and which confirm it, beginning with that long eclipse of the intellect which, particularly as concerns the questions to be examined by us, extends beyond the period of the Middle Ages.

The mathematical portions of the *Theaetetus* and the *Meno* show, however, that the study of the so-called irrational magnitudes was by no means condemned to remain the intellectual scandal which horrified the Pythagoreans. Plato would much rather have sought in it a support for his doctrine of participation in that which attains the sovereign clarity and absolute intelligibility of an Idea—beyond the number itself. In any case the attempt did not have a lasting effect; it was deluged under the confusing multiplicity of aspects in which Platonic teaching had clothed it, or more simply, swept away by the decadence of the ancient Academy.

Not only does the syllogism, considered the universal instrument of reason, henceforth occupy the forefront of the scene, but the authority of Aristotle, who handed down the famous arguments of Zeno of Elea, was to confer upon them a significance and an importance which went far beyond and even contradicted many of their author's intentions. Originally it appears they were intended to refute the Pythagorean hypothesis of a continuum composed of discrete units such as mathematical points. The hope of seizing upon an absolute element vanishes as soon as the process of *dichotomy* comes into play, and the series of indefinitely decreasing values resulting from it— $\frac{1}{2}, \frac{1}{4}, \frac{1}{8}, \frac{1}{16}$, etc.—are exhibited.

On various occasions Leibniz invokes the example of the summation of a similar series in order to explain how science, whose advocate he is, introduces the infinite into the structure of the finite. On the one hand one possesses knowledge of the total series without having to run through the terms one by one, since the law of its formation is contained in one's "reason." On the other hand he is assured that the sum of the series is indeed equal to unity thanks to his knowledge of the dynamics of thought:

> Equality may be considered an infinitely small inequality, and one can make the inequality approach the equality as closely as one wishes.

But—and this is where the disagreement between ancient and modern thought emerges—exactly that which Leibniz perceived "right side out," so to speak, had no meaning or usefulness for Zeno of Elea except when considered "wrong side out," as the instrument of a negative dialectic, which became so popular that the process of Pythagorean dogmatism very quickly degenerated into a state of human thought that indifferently allowed itself to be convinced of the incapacity to treat motion involving the infinite and the continuum. From this arose the obsession of a "guilty conscience," which was eased only by recourse to the roundabout way of the method of exhaustion. Nearly 2000 years are to pass by before, in Gaston Milhaud's words, "the infinitesimal moment of all change," in space as well as in time, will be endowed with a direct expression, in which integration will be understood as a normal operation of thought.

The history of the 17th century shows how gradually these new perspectives were arrived at. Brief as our examination must be, we should keep three names in mind, Descartes, Pascal and Leibniz.

Descartes opens the era of modern mathematics because it is he who introduces an awareness of relationship—or, as Gino Loria felicitously observes, it is due to him that the *verb* is substituted for the *noun* upon which the ancients had fixed their imagination. Algebra, self-generating through its "long chains of reasoning," which the mind unfolds upon its own initiative, is "the key to all the other sciences"; but at the same time Descartes comes to maintain that algebra marks the limits of human capacity. One more instance of a dogmatic and dangerous negation appended to a positive triumph. Close upon the algebraist's death, Pascal repeals the limitation; from intuitions about space in the domain of the infinitesimal, he obtains results of such power and complexity that he will seize the opportunity they afford him to challenge all the scholars of Europe. Then, in the next generation, through the simultaneous work of Leibniz and Newton, analysis will take its revenge. It takes hold of Pascalian geometry and provides a version of it that is transparently clear to the intellect.

These successive bursts of light are accompanied by systematic reflections on the methods employed to increase the body of mathematical knowledge. In Descartes's case, intuition concerning evidence —intuition being the last step in regressing from the analytic to the absolutely simple as well as the starting position for the recomposition of the intelligible object—this intuition travels in the opposite direction from the Aristotelian syllogism, which supposes that induction proceeds up to the genus and then down to species and individual. Pascal in turn renews the advance of deductive logic, for which, the impossible task is proposed, that it define all terms and prove all propositions. Nature makes up for reason's deficiencies with the *heart's* feelings. The heart does not need to express itself in clear and distinct ideas in order to impose its irresistible certainties on the human geometry. Leibniz appeals to this sentiment in his *New method by maxima and minima*. Again, the advent of differential calculus, which puts its inventor fantastically ahead of his times, is in his own eyes only a particular case of a universal Characteristic, which would take him backwards and revive the medieval dream of an absolute deduction, as if man were actually in a position to run down the sources of the primary definitions and even prove the very axioms.

Thus, there are as many discoveries associated with gains in positive knowledge as there are mutually exclusive conceptions of method colliding with each other. Pascal and Leibniz seem to us to be working together to force open the doors of mathematical infinity. But is this to be done by pushing beyond the normal resources of reason? Leibniz parts company with Pascal on this fundamental issue. He returns to the path of Cartesian analysis, while Descartes and Pascal find themselves united in their opposition to Leibniz's position that the deductive process is self-sufficient. The two of them have proclaimed the primacy of intuition, even though they otherwise give it a radically different meaning. One final word: Descartes, Pascal,

Leibniz stand out among the greatest philosophers because they understood that the obligation—the heroic obligation—of going after the
truth with one's whole soul excluded all reservation and all compro-
mise. They denied to thought the right to exempt itself in any degree
—by claiming a loftier or more distant goal—from the requirements
which scientific work has the honor of rigorously obeying. If God is
other than a mathematical "proposition," the idea of truth, which is
one and integral, requires a unity of mind and integrity of conscience.
The same genius that a Descartes, a Pascal and a Leibniz manifest in
their creation of the mathematical world is also present when they are
preoccupied with the acceptance of divine reality. This is why if one
hopes to attain a full appreciation of the understanding and love of
religious conceptions developed in the *Méditations*, the *Pensées*, the
Théodicée, one should observe the mathematical career of their
authors and note there their personal style of behavior towards
God.

In Descartes's case the ontological argument (thus named by Kant
who legitimately attacked its dialectical garb) is understandable and
even justifiable if we consider that it limits itself to clarifying the
intuition which takes possession of the infinite directly. The divine
Cogito precedes and commands the human *Cogito* upon which the
innate rationality of knowledge hangs. The God of Pascal, in the twofold mystery of His justice and mercy, has the same relationship to
Descartes's God that infinitesimal geometry has to analytic geometry.
The natural incomprehensibility surrounding paradoxes about infinity
prepares man for the supernatural incomprehensibility of Biblical and
evangelical teachings. Leibniz deliberately takes a position in opposition to his predecessors' theology. In order to assure logical correctness, he revives the a priori proof of the existence of God as he finds it
in Descartes. This divine existence is the necessary consequence of a
conflict among the possibilities whose claims to existence constitute
the world of eternal essences; in turn this divine existence will
guarantee perfect intelligibility of the universe based on the presumption that the predicate is inherent in the subject.

Thus Descartes, Pascal and Leibniz agree in maintaining that the
concern of mathematics is not a system of nature reduced to the utmost
degree of abstraction, as it was for Aristotle and was to be for Comte;
it is the fitting prelude to, and the relevant proof of, a spiritual doctrine wherein the truths of science and of religion will lend each other
mutual support.

From this source arise differences of opinion all the more irreconcilable in that they are concerned not with the solution of a similar problem but with something far deeper, the manner in which the problem is put: Shall it be in terms of intellectual intuition, or the "reason of the *heart*," or formal deduction? It is superfluous to remind anyone to what degree the diversity of consequences intensifies the interest in the different ways of setting out.

These three alternatives have been presented only to save us from *false alternatives*. Mathematicians voluntarily classify themselves into *logical* and *intuitive* thinkers, but it is the characteristic of an exact science that the manner of investigation has no bearing on the value of a discovery. For that matter, when we say *logician* we do not mean *panlogician*, and when we say *intuitive* we do not mean *intuitionist*. In closing this too brief sketch it will suffice to cite as evidence the conversion of Bertrand Russell—once the boldest of realists—to radical nominalism, as well as the question Henri Poincaré asked himself with reference to continuous functions without derivatives: "How can intuition deceive us that much?" Such examples allow us to believe that for 20th-century thought, what matters always is to recognize the fundamental features of specifically mathematical thinking, in whatever new forms of expression each generation may specify, and to maintain these fundamental characteristics.

49

ON THE TIDE OF THE GREAT CURRENTS:
VIEWS ON MATHEMATICAL TRAINING

by Georges Bouligand

> *Let us hope, then, that mathematical educators will not exaggerate the dogmatic aspect of their science; instead, they might draw examples from the history of science to show that the only objective of rigor is "to sanction and legitimate what the intuition has discovered," as Hadamard so aptly put it.*
>
> RENÉ DUGAS (page 17 of this work)

THESE lines from a great arbiter of thought, a mind constantly in contact with original papers stemming from widely varying periods, and with a matchless gift for seizing upon the most fruitful ideas[1] and imparting them to his readers, lead directly to a theme that is always before us and is more vital today than ever—namely, that of formulating mathematics so as to better meet the need of increasing, qualitatively and quantitatively, the participation of students.

Trends that are well known properly attract our attention to measures formulated thirty years ago to meet a need for unity. But to appreciate the invigorating potential of these measures it is sufficient to peruse one or two books, which, in spite of the number of authors involved, are in accord on what constitutes this best approach.[2] Recent acquisitions in knowledge have produced a thorough sifting with a view toward a revised synthesis. But not everything is finished: One important task is being organized in conformity with the views of René Dugas. I propose to discuss it briefly.

(1) In order to win students over to mathematics one cannot be satisfied with a presentation of the finished product. To be sure, it is

---

[1] The index numbers refer to the bibliography to be found at the end of this article [translator's note].

well to take advantage of the very best textbook put out by the perfect publisher—providing any such exists. At all levels, we must start out gradually and provide both psychological and historical supports. A realistic view of our inquiry requires that the psychological supports be limited to a small number of the most noteworthy *psychological facts*. At this point it is a good idea for us to indicate which facts should be selected.

Some of these facts arise out of logical considerations. For example, the defender of a certain position lines up his arguments before an unfriendly audience and finds that they add up to a contradiction. Another jolt—less severe, it is true—occurs when the development of an incipient theory breaks off at the very point where it is about to lead naturally to the most tangible of results. Yet another jolt occurs when a critical mind notes the omission of one or more links in his so-called "proof" of a theorem.

The way in which information is handed out must be carefully considered from the outset. Thus it is preferable to get the students to work things out experimentally so that they may rather quickly move on and correctly deduce theorems involving simple properties that are spontaneously accepted because they are held to be verifiable. In this way one prepares the final stage where only the axiomatic method is appropriate and for this the *choice of axioms*, as well as the *choice of concepts*, will be suggested but not imposed. This is a good opportunity to bring in some history; one can recount how Hilbert and Weyl were confronted at this stage with the problem of determining what constitutes Euclidean geometry.

Deprived of the tool of deduction, the experimental stage will put history to use. For example, one may refer to Thales, who is supposed to have applied his theorem on proportions to obtain the height of the pyramids by using the shadows they cast. (Today a telegraph pole would be simpler.) Thales is likewise supposed to have made use of a circle circumscribed about a rectangle to establish the property of a triangle inscribed in a semicircle. In addition to these early speculations the Pythagoreans ventured upon the notion of equalities, as in

$$1 + 3 + 5 + \cdots + (2n - 1) = n^2$$
$$1 + 2 + 3 + \cdots + n = \tfrac{1}{2}n(n + 1)$$

by resorting to square numbers and triangular numbers, following the familiar principle which gave rise to the theory of figurate numbers;[3] all these things spotlighted the deductive method.

(2) Why did mathematics not stop with these minor achievements and why did it instead continue to grow? History gives a clear answer. At this period equally great enthusiasm had been aroused for another objective. Having accomplished the quadrature of a given rectangle *R* by using the straightedge and compass to construct a square *Q* with area equal to that of *R*, mathematicians believed it was possible to tackle the more or less obvious problem of the quadrature of the circle. At the outset scholars did not perceive the difficulties, which surpassed by far those created by two other contemporary problems, the *trisection of an angle* and the *doubling of a cube*. The latter two problems reduce to two intersections: in the case of the angle trisection, the intersection of a circle and an equilateral hyperbola; in the case of doubling the volume of a cube, the intersection of two parabolas. While the straightedge and compass cannot master these problems, they do at least, in the second and third cases, permit us to locate as many points as are necessary to trace the two aforesaid solution conics with an arbitrary degree of precision. However, to arrive at an exact solution in each of these cases, we would need an infinite number of points. Here, then, is the *role of the infinite*, which, in an attempt using intersections of lines and circles, introduces a graver impediment to the quadrature of the circle;[4] in this case the straightedge and compass no longer suffice to determine a single point of a curve which would correspond to one of the solution conics in the angle-trisection or cube-doubling problems. In fact the quadrature of the circle gets nowhere without an infinitude of ransoms "paid to the customs guard at the border of infinity"; moreover, this guarded threshold *S** marks, besides, the entry to a strange land discovered by the Pythagoreans when they sought to *measure* the diagonal of a square in terms of its side. Whereupon, alerted by geometers to the operations of measurement and the general conditions pertaining thereto, mathematicians came upon a new threshold *S'*. Thanks to its unbounded curiosity mathematics continued on through thresholds *S* and *S'*.

(3) The effect of this double discovery was to separate physics, ϕ, and mathematics, μ, despite their common objectives. In ϕ one is not afraid to make repeated observations on a very fine material wire, exploiting its inextensibility *I* for the purpose of conceiving its "length." And one does not hesitate to attribute a measure *I* to the wire which is independent of any deformation of the wire. On the other hand,

---

* *S* is the initial letter of *seuil*, the French word for threshold [translator's note].

where abstraction flourishes, in order to "rectify an arc" one would have to overcome the difficulty of decomposing its total length, first into numerous small arcs, then into innumerable *micro-arcs*—venturesome *words*, but they do not advance matters at the outset.

Thus in μ we shall have to guard against incorrect concepts (such as the micro-arc). But it is not sufficient to take defensive measures. We have to exploit suggestions from ϕ in addition. After measuring a length of arc with inextensible wire, we proceed to another type of dimension—the *area of a region*. From a homogeneous sheet of paper, *Fp*,* cut out an area; using a balance, we cut a square of practically equivalent mass from *Fp*. Or we proceed to a new stage, finding the volume of a pebble (filling a region in space). We immerse the pebble in a mass of water in a cylindrical test tube and carefully note the increased height of the liquid after this immersion; this takes for granted the existence of a property I', which here is the incompressibility, a realistic fact. We have thus used experimentation to attain more or less precise results; next, the critical mind is put on the alert, though not to reject out of hand "the information picked up." It will agree to a *schematization* of this information, substituting a system of *abstract relationships* among various elements for a group of remembered images. At the instant of abstraction there is no longer wire, nor *Fp*, nor pebble, nor test tube. We are going to think, without visual crutches, about an arc formed in the figure f, a rudimentary definition characterizing such an f, and demand next a second definition providing for additivity and seek to arrive at the length of f (at least in the favorable cases whose class remains to be determined). *The hunt for a concept* is on! But attention must be given to wording: No longer do we say, "the volume of a pebble," but rather, "the volume of a region." In addition, we unify some of our terms: Area becomes a special case of the volume of a region having the form of a right circular cylinder (with altitude equal to unity). Every concept requires a definition, and this is never completely achieved in one fell swoop. In the preceding example we begin by studying regions which will take up a certain finite number of squares on a graph-paper grid of squares, or, if we are dealing with space, the same procedure will furnish a finite number of cubes of a three-dimensional grid. The "current" is beginning to establish itself, and there is already a glimpse of the use to be made of smaller and smaller grids, which are obtained

---

* *Fp* stands for *feuille de papier* (sheet of paper) [translator's note].

by repeatedly dividing each side of a square (or cube) by two to arrive at the succeeding step.

(4) We find ourselves at a new threshold S'', from which we can begin to glimpse a current of *mathematical activity* with its own dynamics, creator of mathematical *models*. To verify the existence of this current we must separate the specific subject matters of physics and mathematics, μ. Now physics poses *problems*; these can be schematized every time by reducing them to a mathematical type μ. It is therefore a question of extracting from a given (unambiguous) category an element x, which is subject to conditions that are themselves clearly formulated.

To the extent that problems are solved—each by the appropriate construction of element x (whether mechanically, graphically or operatorially)—some content of mathematical information will result; this content will be added to the repertory ρ, consisting of the ideas which determined the *Elements* of Euclid three centuries before our era; in this repertory *concepts*, *methods* and *results* range side by side. Moreover, at different stages these will come under attack by recognized methodological progress; this will call for actual reformulations in ρ. Such a ρ, at a certain stage, is the accrued *synthesis*. The successive stages mark the separate phases of the work carried out in a quasi-permanent spirit; these phases are very clearly indicated in Paul Germain's excellent article in the present work.[5]

The aforesaid study reveals that at each epoch the overall evolution of mathematics is the *resultant of two forms of activity*, say activity P and activity S, the first oriented towards *problems P*, the second towards *synthesis S*. Through this turn of events we see the decisive role played by history and by what can be taken as the invariant elements among its thousand aspects.

But far from being independent, these two modes offer *typical interactions* which I have already drawn attention to.[6] Through this continual interplay there has gradually arisen, and then been enriched, some *operatory material* to answer the need of ever more varied problems and to *construct their solution*. This material has (most often) been elaborated for the sole purpose of assuring, in addition, the uniqueness of groupings of problems under the organizing aegis of a unifying method, groupings which have aided the work of synthesis.

(5) The act of "construction" has just been shown to reflect a desire for efficacy, which is inherent in mathematics considered as a

tool. "Problem grouping" reflects the desire to bring about a good synthesis capable of endowing mathematics both with the simplicity and generality that come with dealing always with very general *concepts* and with the resulting distillation of *axioms*. To broaden means, in effect, to reject supplementary hypotheses, and this in turn means reducing the collection of premises and, beyond this, reducing the number of truly distinct types in every group of problems playing the role of *special cases*.

After much reflection on the typical situation (*RTS* for brevity), the search for simplicity has resulted in our establishing a distinction between those belonging to a class certain of whose representatives at least (*observed RTS, statistical RTS*) immediately strike the mind, and those special cases which owe their development to some scholar's genius for Pascalian finesse. We are talking about *structural RTS*. What these are becomes clear after a comparison of numerous examples reveals certain *recurrences*. The architecture of these recurrences indicates that there is a fixed type of repetition from one row to the next (such a recurrence has a conspicuous place in the theory of figurate numbers). The suggested term *recurring structure* is justified by the many applications in combinatorial algebra (arithmetic triangle). And in the theory of functions we have seen the appearance of many other structures, as for example, the *structure of a group*, which Lagrange created to solve special equations by means of groups of permutations. Galois completely generalized the structure of a group for this same field; it then united Euclidean geometry and its variants,[7] and from there it gradually infiltrated all branches of mathematics. It is worthwhile to recall a prominent example, the *relations of divisibility*, the methodical study of which calls for the following structures: *ring*, derived from the group by supplementary conditions; *field*, derived in similar fashion from the ring; and still others (*ideal, lattice*), all sharing the common property that they are attributable to a set, which by virtue of the appropriate axioms possesses one or another of the above structures (or, if necessary, one from among a wider group).

An equally tangible example, *elementary geometry*, introduced in the plane with a view to isolating the relationships R, P (R = rectilinearity, P = parallelism), in order to derive a "thought experiment" which begins from a slab Δ^* of parallelograms. From the relationships R, P which are implied in Δ, other relationships R, P are developed:

---

* From D, the initial letter of *dalle*, the French word for paving slab [translator's note].

From the former involving vertices and sides, there result others involving vertices and diagonals, which permit one to go from Δ to a smaller Δ′ (four Δ′ paving blocks make up one Δ); then from Δ′ to Δ″ . . . . In the preparatory phase $p\phi$ this method succeeds in isolating the *concepts* which permit one to reconstitute a similar grid *in abstracto*, without figures (in going from binary numbers to real numbers here we need, besides number, the *free vector*, the *vectorial space VE* from which we get the vectors to be treated, the dimension of the aforesaid space, and finally the point), and permit one to formulate the axioms with a view to being able to combine the vectors of *VE* or to determine the dimensions (here $d = 2$), and to specify the interdependence of *vectors* and *points*. Always present in $p\phi$ is the objectivity of the notion, *relation of areas*, between a domain uniting a finite number of parallelograms of Δ, Δ′, Δ″, . . ., and one of these forming Δ, a number that can then be attained without introducing perpendicularity. We see the objectivity of an axiomatics conceived in this manner and made up of substances which are at the same time the chief elements at the basis of vectorial analysis reduced to its linear operations. Once this is established, indeed completed by the *requirements of continuity* (here partaking of the topological), one can rediscover plane geometry, limited to R, P (or affine), by way of deduction, and one can introduce the group of point transformations, conserving the relationship R, P with its various subgroups: translations, similarities, groups invariant with respect to area. Similarly, the path which leads to *metric relationships* (perpendicularity, angles, distances) appears after one has introduced squares in the category of parallelograms, and derived from the square the regular convex polygons with 8, 16, . . ., 2^n sides—(new testimony to the fruitfulness of the binary system.[8] Once geometry has set out on this path, we can appreciate not only the advantages it offers in constructing a synthesis, but even more, the superior mastery it provides over problems.

Thus we see developments which can immediately reinforce teaching at every stage. In this manner the student actually participates in rediscovering and formulating something.

(6) To sum up, by borrowing from epistemological history one finally attains the paths that promote *rediscovery*. In a very natural way, the effect of such remarks is to put in the forefront *invention* and the book Hadamard has devoted to this subject, presenting a wide survey of its connection with *psychology*.[9]

Now, I prefer to use only *psychological facts* here (cf. no. 1). I am selecting two for the discussion that follows:

ψ_1, the malaise that comes from forgetting a link in a deductive scheme;

ψ_2, the malaise experienced when one is blocked in developing an attractive theory which is on the verge of yielding results of real interest.

The choice of ψ_1 and ψ_2 was suggested to me by a theme in Louis Couturat's thesis *The Logic of Leibniz* (Paris, 1901), and by the twin viewpoints presented in that study, according to which the master of algorithms envisioned a *doctrine of proof*, on the one hand, and a *doctrine of invention* on the other; he did this because of his desire for a *universal method* based on a permanent symbolism. In the deductive as well as in the creative areas, he wished, toward this end, to reduce combinations of ideas to combinations of their symbols, and mental operations to operations of a mathematical type. To master the whole field of mathematizable knowledge in this manner imposed a special choice of notation (such a choice led to the infinitesimal calculus). The interest in his symbolism grew as a result of the role it gave to *binary numeration* and explains the *long-term influence of Leibniz's program*. At the present time the power of this program is embodied in the concept of the computer, which is about ready to undertake the task of sorting the true from the false. In addition, beyond the realm of number, Leibniz conceived a *geometrical calculus* for studying the properties of position directly, and in addition a systematic study of logical operations.

Let us grant the principle of a universal method that is, along with logic and mathematics, remarkably well qualified to dominate other doctrines; this method must accept facts ψ_1 and ψ_2, and (without decreeing them a priori) undertake to clarify their role. At the outset it will analyze the conditions under which they obtain. Now, countering the "malaise" which has to be "treated" each time, there appear these two efficacious new facts Ψ_1 and Ψ_2 (each coupled respectively with its precedent facts ψ_1 and ψ_2).

A student whose notes are incomplete discovers a gap in the deductive scheme that he expected would establish a known but rather well hidden result. He often makes the best of a bad job, after writing a letter asking for assistance on one or two specific points. With enough effort he succeeds in filling the gap in sufficient time to avoid sending off the letter. Moreover, in justifying their results the great innovators hardly give us more than incomplete schemes; the attentive reader manages to complete the proof most of the time, by working with the steps that are furnished. (Exceptions exist, nevertheless; witness Fermat's last theorem, pp. 81–91 of Vol. I.)

Framing an examination which would lead candidates to little-known results (for their level) sometimes fails unexpectedly. First the author, in sketching the route to be taken, in the belief that he is letting the students enjoy their initiative, does not allow them enough time. Secondly, it fails by trying for an element of novelty.

Let us now, in a Leibnizian spirit, accept the idea that Ψ_1 answers the difficulty ψ_1, which arises from a gap in a *theoretical scheme*, and that Ψ_2 is the answer to the predicament ψ_2, the case where an *extension* of a theoretical scheme is indicated. Moreover, we are not to view the pairs (ψ_1, Ψ_1) and (ψ_2, Ψ_2) as symbols of schemes whose powers are limited to the deductive area for the former, and the creative area for the second. The two types of *intellectual effort* required by each pair respectively—both demanding original mathematical work—dupli-cate each other rather closely, even if an effort of invention takes the place of a simple effort of memory. Henri Bergson made a surprising observation in a study of *intellectual effort* appearing in the January 1902 issue of the *Revue Philosophique*. According to him, this effort assumes various forms, each of which comes into play at the appropriate level. There is an ascending gradation, in the order of appearance, from the effort of memory to the effort aimed at understanding (or interpreting) and finally to the effort of creative imagination. The fact that inter-mediate steps intervene between these terms does not rule out their common nature. Such a judgment might appear risky were it not confirmed on various grounds. Firstly, the second term of the Berg-sonian gradation is to the third as rediscovery is to authentic discovery. And furthermore, the very aim of an effort of memory is to rediscover what has been experienced. Therefore, comparing the effort of memory and the effort of invention, one can expect to encounter similarities. Now, the legitimacy of this is already evident from the following two observations:

(i) The ability to complete a gap in a theoretical system (fact Ψ_1) by an aptitude for deduction is very similar to the ability to remember a melodic line, thanks to which one recovers the melody in fragments until it is entirely filled in;

(ii) The ability, through an aptitude for invention, to extend an embryonic theory with a view to obtaining worthwhile developments is also very similar to giving body to a musical theme by means of variations, which start out with tame embellishments, then become gradually bolder, and finally go beyond the confines of what it had been reasonably supposed was fixed.

Let us add to the end of Chapter II of Hadamard's book the related idea that the unconscious, which is often an aid to memory, is no less helpful when clarity is wanting, as when a somewhat baffling text is being read; these are phenomena akin to the inspiration which a researcher enjoys from time to time. With respect to the question of the association of ideas, one can conclude, finally, that mechanisms exist capable of buoying up intellectual effort, and that these mechanisms act in the various ascending stages of Bergson's gradations.[10]

(7) After this long digression—too long, but justified by present educational needs—we have a better understanding of how rediscovery may be used to approach new results in a fresh way. The *dualism—problems–synthesis*—which we have arrived at through our examination of the commonest invariants appearing throughout the development of mathematics becomes, in fact, the basis for founding a *rational heuristics*.[11] Furthermore the study of the pairs (ψ_1, Ψ_1) and (ψ_2, Ψ_2) has accented the benefits to be gained from devoting extra attention: in fact recollected psychological episodes show the *benefit of a suddenly redoubled effort* made by a student struggling with his problem. Hence the value of every factor calculated to produce the employment of such an effort. But how does one recognize what measures promote this?

In this matter history can again instruct us. After some timid steps in the direction of hypothetical-deductive systems, closely akin to Euclidean geometry through the instrumentality of the projective group, the role which *adequate models* could be expected to play in analyzing various systems of mathematical-logical relationships was established. Today this method has lost none of its value. Technical advances have sometimes increased it; one sees this particularly with respect to topics on *geometric laminas*,[12] the study of which is linked to kinematics. Now one of Leonhard Euler's justly finest achievements consisted in *founding the general kinematics of continuous media*.[13] Thus in the very difficult research on the non-linear, modern analysis, with Darboux, Poincaré, Hadamard, Cartan and their numerous disciples, is indebted to Euler for the only Ariadne's thread available for this confusing labyrinth; the unique possibility of obtaining a synthesis is the *theory of the moving trihedral*, clad in all its topological finery.

(8) But there are other measures bringing us back to psychology. Leibniz had taken the first steps in this direction when he sought to introduce a program for discovering the new by nearly infallible methods, and by the same means avoiding the expenditure of vain efforts. One may learn more about this in the previously cited thesis by Couturat (page 177).

The twofold program of Leibniz called for two opposite types of processes, one "going from principles to consequences, or from *causes* to *effects*" and the other "going from given consequences to sought-for principles, or from *known effects* to *unknown causes*." This led Leibniz to declare: "The art of inventing is as much synthetic as analytic."

My research in *direct infinitesimal geometry*, beginning in 1928, came to suggest that as far as I was concerned I could without question re-assert the equivalence of principles and causes and the equivalence of conclusions and effects; this is the main point of the article I wrote in 1944 for the present collection on the topic "Intuitive Approaches Toward Some Vital Organs of Mathematics." This article introduces us to the widely applicable guidance offered by the completely Leibnizian idea of *causality* and the companion idea of *stability* (playing the role of generalized continuity in Fréchet's sense)—at least in the area of mathematical creation. I had introduced these ideas between 1932 and 1935 with a view to illuminating the thorough study of existing mathematical theories rather than to provide a guide for research efforts. Thus I was keeping to the second level of the Bergson scale (no. 6). It might be noted in passing that I stayed at this same level in my *Methodological Basis*, published in 1950,[14] and that, in consideration of the extent and difficulty of the task, I allowed several years' delay in leaving this stage, which I did about 1956.[15]

(9) For there were two objectives I needed to attain:

(O_1) To show how the support provided by the dual scheme of *problems–synthesis* (PS) permits the researcher possessing a little knowledge to orient himself in choosing his themes by taking advantage of opportunities (correcting abnormal situations).

(O_2) To show what is suitable for guiding the inexperienced research man who has been given a suggested topic.

Some brief remarks about O_1 first. The program indicated by *PS*, which is designed to give really close attention to the history of mathematical concepts, methods and results, reveals that there are interruptions in the great currents after which they resume again. With reference to the work of Leibniz, I spoke in no. 6 of his long-term influence, testimony to which are: computers, in the role of machines to determine the logical value (true or false) of a more or less plausible statement; the *algebra of logic* created by Boole in 1847; and geometric calculus founded at nearly the same time by Grassman. No. 7 justifies analogous remarks on the work of Euler. Among other examples, one could add the long wait between Desargues–Pascal in the 17th century and Poncelet–Chasles in the 19th, apropos of projective geometry, the latter two coming along to feed the geometric current dominated by group theory.

As for departures of a heterogeneous nature, I shall again cite the *retroactions* appearing throughout history that result from *returning to principles*. Leibniz's retroactive act consisted in founding infinitesimal calculus, Cauchy's in arriving at a satisfactory theory of limits and series. There, too, examples abound, and for this reason we rightly insist on generalizations of the idea of number.

For our purposes I shall consider only two types of situations similar to those just described.

(H_1) The departures of a heterogeneous nature under consideration here became evident around the beginning of my teaching at Poitiers; Differential geometry, which had retained if not 19th-century methods at least variations of these, instead of following the example of the theory of functions and adopting the concepts of set theory, was strengthened by recourse to the moving trihedral on the one hand, and to vectorial analysis on the other. Hence my book of 1932.[16]

(H_2) Another departure of a heterogeneous nature: the presence, in the synthesis, of substantial developments on topics which no grouping of problems could previously make productive. This observation has led to my current work on a grouping of problems—a completely natural grouping from the physical point of view—but *not of a unitary type* because of the absence of a mathematical method capable of handling all aspects of the problems. We are discussing transformations conserving volume, volume being taken where necessary with some generality, borrowing from sets and the contributions which may result from the (closely connected) study of turbulence, through sufficient research on vector fields and on the

Navier-Stokes equations for liquids.[17] In this connection a whole chapter remains to be written on historical recourses to the departures of a heterogeneous nature mentioned.

(10) Let us pass on to objective O_2. The beginner must become accustomed to using *heuristic rules*. He should start with rules that are valid for every topic and continue his research with rules that are gradually limited to those allowing him to approach the topic being studied and to achieve specific results.

I shall start off by recalling the *rule of causality*, which is taken directly from the Leibnizian concept of two contrasting procedures (see no. 8). The ends it aims for are a plausible conclusion and the elimination of subordinate hypotheses. Although the *rule of stability* derives from the same concept, it comes into play only on occasion when recourse is had to an arbitrarily close neighborhood, this restriction applying as much to the premises as to the conclusion; for example, the solution of equations in integers cannot take advantage of this rule at all. To compensate for this, the *rule of constructivity*, which proclaims a privileged position for *everything that can be constructed*, is independent of this last-mentioned rule.

The practical interest of such rules resides in the opportunities they provide for a thorough discussion of how to handle a topic that has been introduced by prepared questions. One cannot state a priori that the rules will permit mastery of a specific point being studied; usually they will involve the kind of progress that is related to the student's improved understanding of the subject, assuming he has sufficient background. Leibniz's claim that he imparted nearly infallible procedures appears excessive.

However, having noted the close relationship between physics and mathematics (nos. 3–4) and having spelled out the relationships between these two fundamental branches of knowledge, we should give priority to the rules that stand ready to promote an even deeper understanding of physical mathematics. This would certainly be the case for the rule of stability (see final remarks, Vol. 1, p. 64, no. 6); this rule was already formulated in very similar terms by Pierre Duhem in 1907 (*La Théorie Physique*, 2nd part, Ch. 3).

The same type of situation is encountered when we examine the conditions under which one goes from the *local* to the *general*. First, as in the above discussion, the local itself comes into play only through neighborhoods. Now physics is still interested in such an examination; in fact, for the experimenter, the effectiveness of the said neighborhoods is specified by an order of magnitude. In particular, the only practical value of the concept of simultaneity is purely local.

All this involves us in incidences of this or that rule operating upon this or that example (a construction achieves its objective only if its principle is stable) and in addition involves other physical specifications such as the impossibility of making a field theory susceptible to experimentation without requiring space-time to verify certain local conditions, in particular to verify admitting such differences in measuring standards. In this respect the mathematics of sets imitates physics; the sets that it considers are not finally determined until the moment when the axioms of the theory being formulated have conferred a particular structure upon them; paradoxical situations are thus eliminated, and the very danger of these seems to impose the aforesaid precaution.

It would be imprudent to try to be exhaustive in the matter of heuristic rules, particularly if one takes those that are to be gradually adapted to a specific topic; these are things to be worked out.

In closing I mention the *statistical rule*, which calls for a *measuring procedure* designed to limit the cases of imprecision of a property P that is apt to escape in a larger context than that for which P has been legitimatized in the beginning. For example, a directed plane curve without multiple points which is cut by a straight line in more than p points (p being given) has everywhere an anterior one-sided tangent α and a posterior one-sided tangent π. Moreover it is established that such an arc A has *bounded length*. In now making this last property a hypothesis, the statement claiming the unique existence of a one-sided tangent α and a one-sided tangent π can be in error; but then, the exceptions arise only in points of A which can be included in a series of partial arcs of A, the sum of whose lengths is arbitrarily small. In other words, the exception is localized in a set of zero measure.

(11) To conclude, I hope that I have shown that this kind of study is truly instructive for anyone destined for research. But its interest goes much further. The orientation that it imparts to teaching from the very beginning makes teaching much more natural and as a result increases its effectiveness.

Today a current is developing in this direction. As evidence I have here the valuable collection of articles on heuristics that appeared in 1958 in another fine book, *La Méthode dans les sciences modernes*, yet another product of François Le Lionnais's initiative.[18] There is a new spirit there which, in the present state of affairs, can contribute to achieving a better balance between problems and synthesis, the two constituents of mathematical activity.

BIBLIOGRAPHY

1. DUGAS, R.: *Histoire de la mécanique*, 1950; *La mécanique au XVIIe siècle*, 1954; *La théorie physique au sens de Boltzmann*, 1959. These three books, published by Le Griffon in Neufchâtel and Dunod in Paris, have highly appreciative forewords by Louis de Broglie.
2. CARTAN, H.; CHOQUET, G.; DIXMIER, J.; DUBREIL, P.; GODEMENT, R.; LELONG, P.; LESIEUR, L.; LICHNÉROWICZ, A.; PISOT, C.; REVUZ, A.; SCHWARTZ, L.; SERRE, J.-P.: *Structures algébriques, structures topologiques*, L'Enseignement Mathématique, Geneva, 1958. *Mathématiques nouvelles*, O.E.C.E., Coll. Royaumont, 1961, with list of manuals.
3. REYMOND, A.: *Histoire des sciences dans l'antiquité gréco-romaine*, Presses Universitaires de France, Paris, pp. 35, 133, 155.
4. LEBESGUE, H.: *Leçons sur les constructions géométriques*, Gauthier-Villars, Paris, 1950.
5. GERMAIN, P.: "A General View of the Evolution of Mathematics," this work, Vol. I, pp. 231–248.
6. BOULIGAND, G.: "L'activité mathématique et son dualisme," *Dialecta* XI, 1957.
7. BOULIGAND, G.: *Les principes de l'analyse géométrique*, Vuibert, 1949, Vol. I, notes III to V.
8. BOULIGAND, G.: *L'accès aux principes de la géométrie euclidienne*, Vuibert, 1951.
9. HADAMARD, J.: *Essai sur la psychologie de l'invention dans le domaine mathématique*, A. Blanchard, 1959.
10. BERGSON, H.: "L'effort intellectuel," reprinted in *L'Energie spirituelle*, Alcan, 1920.
11. BOULIGAND, G.: *Comptes rendus de l'Académie des Sciences*, 252, Jan. 4, 1961, pp. 37–40. "Aptitude à prévoir et mathématique en expansion," *Revue de Synthèse*, July–Dec. 1960.
12. BOULIGAND, G.: "Thèmes géométriques filmables," *Rev. Math. Spéc. Vuibert*, May 1958.
13. BOULIGAND, G.: "L'œuvre d'Euler et la mécanique des fluides au XVIIIe siècle," *Rev. Hist. Sc. et appl.*, XIII, 1960, pp. 113–115.
14. BOULIGAND, G.: *Base méthodologique* (Vol. IIA of *Les principes de l'analyse géométrique*), 1950, pp. 82–85 and 189–196.
15. BOULIGAND, G.: *Le repérage de la pensée mathématique et ses applications à la recherche*, Conf. Pal. Déc. A, No. 223, 1956. *Aspects de la mathématisation*, same series, No. 242.

16. BOULIGAND, G.: *Introduction à la géométrie infinitésimal directe*, Vuibert, 1932, with a foreword by Cartan.

17. BOULIGAND, G.: *La pensée mathématisante*, Vrin, in preparation; 1961 notice in *Titres et travaux*, Berger-Levrault, Section III(c).

18. LE LIONNAIS, F., ed.: *La méthode dans les sciences modernes*, special number of *Travail et Méthode*, Editions Science et Industrie, Paris, 1958.

50

MODERN AXIOMATIC METHODS
AND THE FOUNDATIONS OF
MATHEMATICS

by Jean Dieudonné

IF we are to speak of a "Greek miracle" in the evolution of human thought, surely we are to see it in the astonishing birth of a mode of thought practically nonexistent prior to the Greeks, and brought by them, in a single stroke, to a perfection unequaled for 20 centuries: *deductive reasoning*, i.e., a chain of propositions arranged in such a way that the reader (or auditor) sees himself *compelled* to consider each of the links to be true, once he has admitted the truth of those that precede it in the reasoning.

It is this discovery which gives rise to mathematics as a pure science, since originally it was only a branch of dialectics, differentiated merely by the particular nature of the entities—geometric figures and numbers—upon which the reasoning bore.

Now, the essential condition in applying deductive reasoning is the truth of the premises. When mathematics was established as an autonomous science and when the Greeks began to codify its propositions, they naturally arrived at *primary* propositions, that is, propositions whose truth had to be admitted *without proof*. In Euclid some of these propositions are called *axioms*, others *postulates*. Essentially the axioms are propositions concerning general notions centering around "magnitude"—"larger," "equal," "all," "part" (for example, "two things equal to a third are equal to each other," or "the whole is larger than its parts"); they are to all appearances considered *evident* because they are necessarily involved in the mental picture that we can have of these notions. Postulates deal with mathematical entities proper; their truth seems to come rather from an extrapolation of experiential truths, as for example, the postulate stating that every straight line can be extended indefinitely, or the one granting the equality of all right angles.

As for the famous postulate on parallels, we know that the evidence for it is considerably less strong, so that from ancient times on mathematicians have striven to prove it; it may be that it was accepted as a postulate on the strength of more or less crude experimental verifications of certain of its consequences. However that may be, from the moment that postulates and axioms are regarded as truths, it is particularly necessary to state them in a language whose terms are all perfectly clear. Certain of these terms can be readily defined on the basis of others; these are the ones that are introduced, in short, only to abbreviate the language. But there remain indefinable terms such as point, straight line and plane; the preceding remark is valid for these, in any case. We note, however, that Euclid strives to give definitions of these, and that some of them are more obscure than the notions they are meant to define (for example, "A straight line is a line which is similar to itself at all points"); others are purely negative (for example, "A point is that which has no parts"). It appears therefore that these notions are not as clear as they seem to be; and in fact Euclid's negative definitions are bars against such interpretations as that of Pythagoreans, which endowed points with extension and which had to be abandoned after the discovery of irrationals.

To sum up, the situation in mathematics at the time of Euclid and up to the 19th century is as follows: One reasons about notions of which he has a rather vague idea—they are thought of as a kind of idealization of experiential notions; we admit a certain number of true propositions about them which also appear to be extrapolations from experience. Even when this situation alters for the worse, as after the introduction of infinitely small and imaginary quantities, with the interminable discussions that the problem of their "nature" raised, we are scarcely ruffled; the reason is that conclusions of deductive reasoning, like axioms themselves, continue to be intuitive in nature and close to experimental facts; moreover the application of mathematics to the experimental sciences, far from leading to absurdities, results in a new upsurge of these sciences and causes them to advance from success to success. With the ends justifying the means, mathematicians thus forge ahead, developing their subject further and not worrying too much about the foundations on which it rests.

This attitude is modified by mathematical concepts making their appearance with advances in analysis in the second half of the 19th century. Contemporary mathematicians were dumbfounded by these concepts; curves without tangents, curves which fill a square, non-ruled surfaces applicable on a plane—first specimens in a gallery of monsters that has not stopped growing to this very day. It therefore appears indisputable that the extrapolation that led from experimental notions to mathematical notions is far from being as natural and as harmless an operation as had been believed up to then; for the first time, moreover, one learns to distrust intuition in mathematical reasoning, since facts as intuitive as the existence of a tangent to a curve are generally mathematically incorrect.

Hence the absolute necessity from now on for every mathematician concerned with intellectual probity to present his reasonings in *axiomatic* form, i.e., in a form where propositions are linked *by virtue of rules of logic only*, all intuitive "evidence" which may suggest expressions to the mind being deliberately disregarded. We are saying that this is a form imposed on the *presentation* of the results; but this does not lessen in any way the role of intuition in their *discovery*. Among the majority of researchers the role of intuition is considerable, and no matter how confused it may be, an intuition about the mathematical phenomena being studied often puts them on the track leading to their goal. But the terrain which intuition thus conquers in a single leap remains to be organized afterwards; the chain of propositions which will end in the sought-for result remains to be built up, link by link. And in this work, intuition is not to occupy the smallest role; strict logic alone reigns, and it is in its cold light that the truths must be examined which the mathematician flatters himself he has already discovered. It is ungrateful and often painstaking work, but how very useful, for whoever has engaged in mathematical research knows that a correct intuition rarely reveals itself right off, and that the largest part of his labors consists in disposing of one false intuition after another! Some eminent minds have long refused to submit to this kind of discipline, preferring to accept the risk of letting the suspicion of inexactitude hover over their work. To cite only one of the greatest of these, everyone knows that Henri Poincaré's works include numerous errors, minor, it is true, and in no way tarnishing his glory; but who does not see that if this attitude were made general and came to be adopted by mathematicians of lesser intellect, that would be the end of the privileged position mathematics occupies among the sciences, and the end of the unique property that it enjoys of *compelling* the reader's assent, since then every mathematical proposition would be as debatable as the intuition of its author!

Axiomatic methods are often criticized for their dryness and sterility. Insofar as the first point is concerned, everything depends essentially on the writer's talent for exposition; nothing prevents him, while remaining perfectly rigorous, from choosing language sufficiently vivid to arouse appropriate intuitive resonances on the part of the reader. As for the second objection, the history of the development of mathematics over the past thirty years is sufficient to make naught of it. The use of the axiomatic method, by showing clearly the source of each proposition and by showing which were the essential hypotheses and the superfluous hypotheses, has revealed unsuspected analogies and permitted extended generalizations; the origin of the modern developments of algebra, topology and group theory is to be found only in the employment of axiomatic methods.

Once the necessity of the axiomatic method was recognized, what were to be the new bases of the science of mathematics? In the first place, the task of revising previously developed theories proved indispensable; for it was apparent that intuitive considerations had slipped into every step of the previous reasonings, and that the so-called system of axioms at the base of these reasonings was not adequate to develop them with complete rigor. The most celebrated of these revisions (doubtless because it concerned that part of mathematics, familiar to the largest number of people, which had until that time passed for a model of rigor) was the revision of Euclidean geometry that Hilbert made in his *Grundlagen der Geometrie*, which appeared in 1899. In it he formulated a system of 21 axioms and showed that these axioms were *necessary* and *sufficient* to prove rigorously all of the known propositions of two- and three-dimensional Euclidean geometry.

The statements of these axioms retained their traditional form, i.e., they were concerned with points, straight lines and planes. But since introducing into the reasoning intuitive properties habitually associated with the entities designated by those words was outlawed once and for all, it became clear that it was quite pointless (at least for the purpose of comprehending proofs) for these words to designate precise notions. It was sufficient (and this was the point of view of Hilbert himself) to consider them as names for three kinds of entities about whose nature one did not have to make any hypotheses; but it was agreed that there was a *relationship* among them which the axioms expressed. Thus it was seen to be perfectly clear that mathematics is the logical study of *relationships* among certain entities, and not the study of their *nature*. So one was free to replace the words "point," "straight line," "plane," etc. in a statement with any others whatsoever, provided one did not modify in any way the relationships into which these words entered. If one then arrived at statements still having intuitive sense, one had a new intuitive *interpretation* of the propositions of geometry. Thus by translating the propositions of Euclidean geometry relating to a *pencil* (straight lines and planes passing through a single point), with the aid of the following dictionary, we obtain the propositions of one of the two-dimensional *non-Euclidean geometries*, that of Riemann:

pencil	plane
straight line of the pencil	point of the plane
plane of the pencil	straight line of the plane
angle between two straight lines	distance between two points
dihedral angle	angle between two straight lines
trihedral	triangle

An analogous interpretation of Lobachevsky's geometry can be obtained by starting with propositions dealing with semicircles orthogonal to a straight line (Poincaré's half-plane).

In the same period other mathematicians carried out work analogous to Hilbert's in all the other branches of mathematics; one after the other, the various geometries, arithmetic, algebra, group theory, the theory of the functions of real and complex variables were *axiomatized*, always by the same procedure. One took as point of departure a *set* of elements completely indeterminate in nature, but among which certain relationships existed, these relations themselves being subject to certain conditions. The system of these conditions constituted the system of *axioms* of the theory, from which all the other propositions were then to be deduced by means of rules of logic alone. For example, a *group* can be defined as a set of elements for which a function of two variables $f(a, b)$ is given, whose value, no matter what a and b are in the set, is still an element of this set; furthermore, the axioms of the theory are as follows: 1. $f(f(a, b), c) = f(a, f(b, c))$, whatever a, b, c may be; 2. There exists an element e of the set such that $f(e, a) = a$ whatever a may be; 3. Whatever a may be, there exists an element b such that $f(b, a) = e$.

With the conclusion of this tremendous task of axiomatizing

mathematics, what result had been attained? First of all, as far as the *truth* of mathematical propositions was concerned, this had been completely divorced from the notion of *experimental truth*. Admittedly this action caused mathematical truth to lose the *absolute* character which so captivated our ancestors; it was now a question of some sort of *hypothetical* truth, that is to say, mathematical propositions were no longer to be considered true except by virtue of a purely arbitrary decree that declared the axioms to be true; the one remaining *absolute* truth was that of the *rules of logic*.

Secondly, what *meaning* did mathematical language have from this new point of view? Put another way, what mental images were the words that one used supposed to correspond to? Of course there was a point to saying that it was not necessary to have a precise mental image of the objects which were to be reasoned about. As Bertrand Russell said, "Mathematics is the science where one does not know what one is speaking about, nor if what one says is true." But there was at least one quality attributed to these mysterious objects, that of *existence*; and along with *existence*, the property of being elements of a *set*, of having *relationships* among each other, of being in *correspondence* with each other, and finally the property of being objects of deductive reasoning according to the old rules of Aristotelian logic, which had been brought up to date and codified by logicians of the 19th century (Boole, Peano, Frege).

In short, the indefinable notions of the new mathematics, about which it was, however, necessary to have a clear mental image, were the notion of *set* and all its related notions: correspondence, function, subset, sum (union) of sets, etc.; in a word, all the notions constituting the proper study of the youthful theory of sets, just created by the genius of Cantor.

Now the notion of set, or if you wish of *collection*, is indeed perfectly clear when it is applied to *concrete* objects and to small numbers of them. If I speak of the set of people present in the room I am in, or the set of letters printed on this page, I have the feeling that I know exactly what these words mean, and this sentiment is certainly shared by my readers. When I speak of the set of men living at the present second, my notion of it is already much more confused. If I now say the words, "the set of positive integers," do I really have a clear mental image of this? If mathematicians were asked this question, many of them would doubtless answer in the affirmative. But if they sought a precise mental picture, is it in fact certain that they would find something different from the collection of symbols "1, 2, 3, ..., n ...," which they have had so many occasions to write on the blackboard? Or for those whose imagination is more visual, the mental picture of an unending road which is staked out by a string of telegraph poles as far as the eyes can see. Here one confronts the problem of *infinity*; nobody would maintain that it is ever possible for a human being to perceive an infinity of objects; but one is sometimes less categorical when it is a question of the possibility of *imagining* an infinity of objects. To our mind the answer should be just as negative; the so-called notion of infinity is only the *illusion of a notion*. Or, if you wish, it is only an extrapolation from a vague notion of experience, that of a "very large number." And at what point this extrapolation transcends the intuition has been shown by the results of the theory of "powers" erected by Cantor, results so offensive to common sense!

And yet, it was indeed *infinite sets* which thenceforth appeared to be the primary notions of mathematics, and not only infinite sets as "simple" as the set of integers, but sets of any "power" whatsoever. And in the wake of Cantor's school mathematicians from all areas of the subject began to reason about sets of functions, sets of sets, sets of sets of sets, etc. Furthermore, Cantor himself agreed in advance as to the legitimacy of such reasonings since he defined a set as "a collection of definite and distinct objects of our intuition or of our thought."

Now, although these reasonings often led to new and interesting results, a growing sense of uneasiness was nevertheless stirred up in many mathematicians towards the beginning of this century. Had we banished recourse to intuition in the area of elementary geometrical concepts, although these touched reality closely, only to be now obliged to reason in just as intuitive a manner about concepts far more difficult for the mind to grasp? The point that aroused most criticism was the application of logical terms and rules to elements of infinite sets, and particularly the use of the terms "whatever it may be" and "there exists"; some people saw an abuse here for which there was no justification at all. In fact, when I state the proposition, "Among the integers included between 3 and 7, there exists a prime number," this is only another way of saying, "4 is prime, or 5 is prime, or 6 is prime." In a general way, to say that in a set comprising a *fixed number of elements there exists* an element possessing a certain property, is to say that either the first of these elements possesses this property, or the second possesses it, or the third and so on until we have exhausted all the elements of the set (it being understood that this manner of speaking does not exclude several elements of the set from possessing the property in question). This successive singling-out of the elements of the set is naturally no longer possible when there is an infinity of elements. However, the rules of logic concerning the use of the term "there exists" obviously imply such a singling-out. There was therefore good reason to question the validity of these rules in reasonings about infinite sets. To tell the truth, this objection could have been raised much earlier, for "theorems of existence" were nothing new in mathematics. But in all the known instances (d'Alembert's theorem, existence of the integrals of a differential equation), at the same time that the *existence* of a number or a function satisfying certain conditions was proved, there was given a *procedure for constructing* this number or this function (in fact, a law of recurrence determining a series of numbers or functions tending towards the number or the function under consideration). In the classic portions of mathematics there was no example of *pure* theorems of existence, i.e., where nothing could be affirmed about the entity whose existence was being proved apart from the existence itself (it is no longer the same today). The appearance of similar theorems in the theory of sets (notably the famous theorem of Zermelo in 1904) inspired in many mathematicians an unshakable mistrust in the reasoning which led up to it, and as a backlash, in all of Cantor's work.

Unfortunately at about that same time another circumstance occurred which threw discredit upon the theory of sets. This was the discovery of what have been called the *paradoxes* of the theory of sets, which show that Cantor's initial conception was untenable. In fact, Cantor's definitions did not rule out consideration of the *set of all sets* as a mathematical entity, but it is readily shown that the properties of this set are *contradictory*. Contradictions also appeared in other directions, for example, in the theory of ordinal numbers, one of Cantor's most original creations. The theory of sets seemed to be collapsing.

But what was then to become of the foundations of the various mathematical theories, since all of them presupposed, to a more or less great extent, this very theory of sets? Had so many efforts to provide a clear and stable foundation for mathematics proved in vain, and was it all to end in a more disastrous situation than the one it had started from? Would one have to renounce forever the possibility of escaping from doubts and uncertainties in these matters? One could well have believed this, to look at the disarray and the extraordinary

confusion which reigned in this subject among mathematicians during
the first fifteen years of the 20th century. (One can get some idea of
this by reading the later chapters of Fraenkel's book *Einleitung in die
Mengenlehre.*) Everything seemed to have been questioned; everybody
accepted or denied, according to his own tastes and tendencies, the
validity of this or that reasoning, of this or that part of analysis. On all
sides there sprang up mathematical "schools," battling each other
furiously; and even within these schools it would have been difficult to
find two mathematicians in perfect agreement on all the points in
dispute. The initial object of the debate, the famous "paradoxes,"
led to the most divergent conclusions; for some, it was the straw that
broke the camel's back, and they used it as an excuse to reject, *en bloc,*
the largest part of classical analysis, indeed even certain rules of logic!
Others, on the contrary, considered these paradoxes as only a sort of
insignificant play on words, having nothing in common with the use
made *in mathematics* of the theory of sets, and resulting solely from a
tendentious interpretation of Cantor's definition. And between these
two extreme positions were represented all shades of opinion. In short,
never since the discovery of irrationals had mathematics passed
through such a crisis; and never, since the great dispute about indivisibles, had it been the occasion of such a flood of metaphysics.

At the present time the last echoes of this great battle are gradually
dying down, and the questions over which there was so much agitation 25 years ago have lost much of their sharpness and disturbing
character. The reason is that, from the midst of this confused mêlée, a
coherent and stable point of view has managed to emerge, which has
gradually won the support of the majority of the younger generation
of mathematicians. This conception, called *formalist,* is essentially due
to Hilbert, who thus consummated the work of clarification and
reconstruction he had begun twenty years earlier.

Let us return to our original question; it concerns the establishment
of the foundations of mathematics upon *incontrovertible* bases; that is,
the terms figuring in a mathematics text should have the same meaning
for everybody who reads it, and moreover, there should exist an
objective method for distinguishing, beyond the possibility of argument,
whether this text is correct. Now Hilbert's essential thesis, which, as
we have said above, is ours, too, states that clear thinking can exist
only with reference to *determinate* objects of *finite* and *experimental*
number (i.e., such that can be subjected to the operation of *counting*).
To wish, as did the first attempts at axiomatics, to have mathematical

reasoning bear on entities indeterminate and infinite in number, is to court inevitable obscurities and uncertainties, like every effort of thought seeking to pass beyond the limits which the very nature of the human mind imposes on its activity.

But if we are willing to remain in the sphere of the finite, can we still perform most mathematical reasonings? Can we even state most of the propositions and retain meaning? When I state the following proposition concerning real numbers, "For any x and y, $x + y = y + x$," does not this express the fact that for *all* the real values with which I replace x and y, the sums $x + y$ and $y + x$ will be equal? As the axiomatic method teaches, I can indeed disregard intuition altogether in dealing with the characteristic nature of the real numbers and reason about them as I do about indeterminate entities for which only the properties stated in the axioms enter into consideration. But I cannot keep from thinking that my proposition is of an entirely different order from the one I would be formulating if I successively replaced the pair (x, y) in the relationship $x + y = y + x$ by no matter which pairs of fixed real numbers, and then connected the propositions obtained, using the symbol "and." Does not the introduction of a notion of infinity appear unavoidable in statements such as these?

Hilbert escapes this dilemma in a manner as simple as it is original. To be sure, there is an inconsistency between the *content* of the preceding proposition and the point of view that thought must always restrict itself to the determinate; but there is no such inconsistency between this point of view and *writing* this proposition which appears as a collection in a certain order of the symbols $x, y, +, =$ and "no matter which"; these symbols are indeed determined and very small in number. It appears impossible to avoid the notion of infinity as long as one considers that the essence of a proposition is its *content*, i.e., the mental image of which it is the symbol. But the difficulty vanishes if on the contrary one agrees that the essence of a proposition is its *form*, in other words, if one agrees that *it is needless for a proposition to evoke any other mental image than the perception of the symbols used to write it*.

This position is so strange and paradoxical at first encounter that it does not seem even worthy of examination; can one maintain, without going astray, that it is possible to carry on reasoning without *understanding* a single word of what is being said? For after all, what are the words of a language and what are mathematical signs if not *symbols* intended to communicate mental images, and how can one separate the former from these images without making them immediately unintelligible?

And yet, this is Hilbert's fundamental idea: The symbols employed in the statements of mathematical propositions are "purely symbols," devoid of any significance. To see how we can arrive at a similar point of view, let us adopt, as an analogy a *game*, for example the *game of chess*; with the *pieces* of this game, which are of various forms and colors in order that we may distinguish them, we form groupings on the chessboard, which succeed each other in accordance with a procedure that is subject to certain *rules* that are fixed once for all, but of such a sort that when the rules allow several different moves, the choice is left to the discretion of the player. For our purposes, the essential thing to note is that in order to play the game of chess correctly, it is not at all necessary to associate with each move a new mental image, of which the move would be the symbol. It is sufficient to assure oneself that the rules are being carefully observed, and this can be done by noting the move itself.

Let us transpose this to mathematics, and we shall have Hilbert's conception; mathematics becomes a *game*, whose *pieces* are graphic symbols distinguished from each other by their forms; with these symbols we make groupings which will be called *relationships* or *terms* according to their form. By virtue of certain rules, certain relationships are described as *true*; other rules permit the construction of true relationships either from any relationships whatsoever or from other true relationships. The essential point is that these rules are of such a nature that in order to verify that they are being observed, it is sufficient to examine the *form* of the groupings which come into play.

It is important to note that the word "true" has in this context a purely *conventional* meaning only, as do the words "relationship" and "term"; we can compare these words to chess terms such as "check," "checkmate," "castling," which are used to characterize, in abbreviated form, certain situations or moves that the rules describe at greater length. Similarly, the mathematical words we have just used are abbreviations to designate groupings whose make-up is given in specific detail by the rules. In any case, it is no longer a question of *absolute* truth, either of relationships or of the rules which permit one to step from one "true" relationship to another. One may say that according to this conception, mathematics no more invokes metaphysical notions and hypotheses than does, for example, the game of chess. And in the same way that everybody agrees to apply an objective method to recognize whether a move in a chess game is correct, so shall we apply an objective method to recognize whether a page of mathematics is or is not correct. Thus the problem of the foundations of mathematics will be completely solved.

Naturally it remains to be shown that Hilbert's conception is realizable, and this can only be done by actually carrying it out. It is impossible to do so here, even in brief form. We shall only be able to furnish a very rapid glimpse of the general direction of the method followed.

Moreover, there are many equivalent ways of developing a system of formalized mathematics. Here we shall sketch the one selected by Bourbaki, and we refer the reader to this author's *Eléments* (Book One, "Théorie des Ensembles," Chaps. 1 and 2) for a detailed exposition.

A certain number of mathematical symbols (game pieces) are used; some of these are *symbols of abbreviation*, which can always be dispensed with at the price of complicating the writing; for the sake of simplifying the exposition, we shall suppose that these are not used. Mathematical symbols are then two in kind; firstly, there are letters (the choice of which is left to the mathematician's judgment); secondly, the following seven symbols:

$$\Box, \tau, \vee, \neg, =, \in, \supset \qquad (1)$$

Then a certain number of *rules* are stated, consisting of several types:

I. A primary group of rules specifies which groupings formed with the preceding symbols shall be called *relationships* or *terms*; for example, $\in xy$ is a relationship, $\supset xy$ is a term, xy is neither a term nor a relationship. Any grouping whatsoever being given, it is always possible by a quasi-mechanical method to determine whether it is a term, a relationship or neither the one nor the other.

II. A second group of rules gives a list of the relationships which will be called *true*: five among them (called *explicit axioms* of the theory of sets) contain no letter at all; eight other rules are described as "axiomatic schemes"; these rules say that when one applies relationships or terms, whatever they may be, as long as they satisfy certain formal conditions, one can form true relationships with the aid of these terms or relationships by means of an explicit procedure. For example, if R and S represent any relationships *whatsoever*, the grouping

$$\lor \neg R \lor RS \qquad (2)$$

is always a true relationship.

III. Finally, a third rule tells what a *proof* is: It is a succession of relationships each of which must be either a true relationship formed by virtue of the rules in *II*, or a relationship R such that in the proof, *before* R, there are two relationships such that if one is denoted by S, the other is $\lor \neg RS$. A relationship is said to be *true* (or a *theorem* or *proposition*) when it can be inserted in a proof (in the above sense).

In broad outline, this is the system of rules of *formalized* mathematics. From the formalist point of view, there is no obvious reason for choosing these rules rather than others, and one can therefore imagine as many mathematical systems as one wishes. In fact these rules were adopted with a view to reconstructing *existing* mathematics. The seven symbols (1) can be *interpreted* in everyday language (for example $\lor$ signifies "or," $\neg$ signifies "not," $\in$ signifies "belongs to"). When the symbols used in a proposition obtained by following the preceding rules are given their meanings and the letters are interpreted as variables (in the intuitive sense of this term) with values in any sets whatever, then one must obtain propositions with an equally intuitive meaning. When the rules themselves are given this interpretation and when it is further supposed that the "variables" represent collections of explicit objects, then the rules themselves reduce to pure truisms. This sheds a clear light on the character of the extrapolation that gives rise to the mathematics originating in the "givens" of experience.

We would be carried too far astray if we were to develop further the formalist method just sketched and were to show how, by pursuing this method, the whole theory of sets, and consequently all of mathematics, can be completely reconstructed. Let us indicate simply that with this conception the famous "paradoxes" disappear of themselves, some of them because we cannot formalize their statement, others because in formalizing them we produce an *unallowable* concept (for example, it can be shown that in the system above the "set of all sets" does *not* exist).

The method above, although inspired by Hilbert's ideas, does not in fact come from his own work or that of mathematicians of his school. Hilbert and his students brought their efforts to bear particularly on the theory of proof, or *metamathematics*, about which a word remains to be said before we conclude.

If we interpret rule *II* in contemporary terms, it means that if a relationship and its negation are simultaneously true, then *every other* relationship is true; it is clear that if this were so, mathematics would lose all its interest. We say that it would then be *contradictory*. It is natural to ask if this possibility, which so far has never been observed, can or cannot occur; this is what is known as the problem of the *consistency of mathematics*.

If one could write *every* mathematical relationship, the question would admit of an answer on an experimental basis, but it is clear that this constitutes an impossibility of the same order as the impossibility of writing all the integers. One can then adopt two different attitudes with regard to the problem of consistency: The first consists in considering that by reason of the intuitive origin of mathematics the likelihood of the absence of contradiction is very great, as great for example as the likelihood of the sun's daily rising; for in this latter case, one can in the last analysis give no other reason for our belief than the fact that up to now the sun has risen every morning, and the same argument applies to the consistency of mathematics. One may add that if some day it is shown that mathematics is contradictory, it is probable that we shall know which rule to attribute the result to, and that the contradiction will be avoided by leaving this rule out or modifying it suitably. In short, mathematics will take a change in direction but will not disappear as a science. This is not entirely a speculation; it is almost what happened after the discovery of irrationals. Far from deploring it as having revealed a contradiction in Pythagorean mathematics, we consider it today as one of the great victories of the human spirit.

Over against this empirical position Hilbert's attitude is much more ambitious; it aims at nothing less than *proving* the consistency of mathematics. Of course, a *mathematical proof* of this is out of the question; the sentence "Mathematics is consistent," or every other sentence having the same meaning, is not a *mathematical relationship* in the sense with which we have endowed this term (no more for example than the sentence "2 + 2 = 4 is a true relationship" is a relationship; what *is* a relationship is "2 + 2 = 4"). Therefore the sentence "Mathematics is consistent" is not capable of figuring in a proof such as we have described above. It is a question of proof in another deductive science which Hilbert proposes to name *meta-*

mathematics and which can be compared, for example, with un-
formalized arithmetic, i.e., with the subject of integers considered as
objects of the intuition. Similarly metamathematics would be the
science of formalized mathematical *relationships* (i.e., the groupings of
symbols described above) considered as objects of the intuition (in
other words, we would ask to have a mental image of "any relationship whatsoever," however complicated it might be, just as arithmetic supposes a mental image of "any integer whatsoever," however
large it may be). Moreover, nothing would prevent the *formalization*
of this science in its turn, just as arithmetic had been formalized. And,
according to Hilbert, the rules of this new game would be even
simpler and more akin to intuition than mathematical rules are. It is
no less true that one thus opens the way to an indefinite process of
regression, for once the consistency of mathematics was supposed
proved, it would be necessary to "prove" that of metamathematics,
which would fall within the province of a *metametamathematics*, and so
on. Without detracting in any measure from the merit of the meta-
mathematical research of Hilbert's school, which has deepened and
rendered more precise our knowledge of the mechanism of mathematical proof, one can remain properly skeptical as to the philosophical
value of this effort. Moreover, according to the most recent studies it
seems that, contrary to what Hilbert believed, the rules which it
would be necessary to adopt in metamathematics in order to come up
with a proof of the consistency of mathematics would be of as high a
degree of abstraction as the mathematical rules themselves; this
would also reduce any significance that such a "proof" might have.

So we shall say no more about metamathematics, leaving to one
side other questions associated with the problem of consistency (such as
that of the *independence* of mathematical rules, or that of the possibility
of proving with a given system of rules that a specific relationship is
necessarily true or false). It is important only to emphasize that the
formalist method and metamathematics are not of necessity linked to
one another. If there had frequently been a tendency to do this, it is no
doubt because the one was the indispensable complement of the other
in Hilbert's mind. But we have seen above that nothing obliges us to
adopt this point of view.

In conclusion, what judgment shall we pass on Hilbert's axiomatic
and formalist method? We have mentioned above the interest which
the axiomatization of theories holds for technical mathematics; the
rise that such theories as algebra and topology have experienced for a

quarter of a century are due to this axiomatization. The interest is not so great where logic and the theory of "abstract" sets are concerned. For despite the great philosophical disputes raised by the "paradoxes," mathematicians have not ceased to be in basic agreement about the logical value of practically all of their reasonings, without being obliged to present them in the form described above.

But leaving aside all question of technical utility, the principal merit of the formalist method will be to have definitively dissipated the obscurities which still bore down on mathematical thought. To all who cherish clarity and rigor, it brings the immense satisfaction of finally knowing what the nature of their science is, of being able finally to reason in full light, of escaping finally the grievous obligation of concealing by verbiage and qualifying phrases the vague and ephemeral character of fundamental mathematical notions. One can therefore in all justice say that it will have advanced science, for, in Jacobi's proud words, "The goal of science is the honor of the human spirit."

A CATALOGUE OF SELECTED DOVER BOOKS
IN ALL FIELDS OF INTEREST

A CATALOGUE OF SELECTED DOVER BOOKS
IN ALL FIELDS OF INTEREST

AMERICA'S OLD MASTERS, James T. Flexner. Four men emerged unexpectedly from provincial 18th century America to leadership in European art: Benjamin West, J. S. Copley, C. R. Peale, Gilbert Stuart. Brilliant coverage of lives and contributions. Revised, 1967 edition. 69 plates. 365pp. of text.

21806-6 Paperbound $2.75

FIRST FLOWERS OF OUR WILDERNESS: AMERICAN PAINTING, THE COLONIAL PERIOD, James T. Flexner. Painters, and regional painting traditions from earliest Colonial times up to the emergence of Copley, West and Peale Sr., Foster, Gustavus Hesselius, Feke, John Smibert and many anonymous painters in the primitive manner. Engaging presentation, with 162 illustrations. xxii + 368pp.

22180-6 Paperbound $3.50

THE LIGHT OF DISTANT SKIES: AMERICAN PAINTING, 1760-1835, James T. Flexner. The great generation of early American painters goes to Europe to learn and to teach: West, Copley, Gilbert Stuart and others. Allston, Trumbull, Morse; also contemporary American painters—primitives, derivatives, academics—who remained in America. 102 illustrations. xiii + 306pp. 22179-2 Paperbound $3.00

A HISTORY OF THE RISE AND PROGRESS OF THE ARTS OF DESIGN IN THE UNITED STATES, William Dunlap. Much the richest mine of information on early American painters, sculptors, architects, engravers, miniaturists, etc. The only source of information for scores of artists, the major primary source for many others. Unabridged reprint of rare original 1834 edition, with new introduction by James T. Flexner, and 394 new illustrations. Edited by Rita Weiss. 6⅝ x 9⅝.

21695-0, 21696-9, 21697-7 Three volumes, Paperbound $13.50

EPOCHS OF CHINESE AND JAPANESE ART, Ernest F. Fenollosa. From primitive Chinese art to the 20th century, thorough history, explanation of every important art period and form, including Japanese woodcuts; main stress on China and Japan, but Tibet, Korea also included. Still unexcelled for its detailed, rich coverage of cultural background, aesthetic elements, diffusion studies, particularly of the historical period. 2nd, 1913 edition. 242 illustrations. lii + 439pp. of text.

20364-6, 20365-4 Two volumes, Paperbound $5.00

THE GENTLE ART OF MAKING ENEMIES, James A. M. Whistler. Greatest wit of his day deflates Oscar Wilde, Ruskin, Swinburne; strikes back at inane critics, exhibitions, art journalism; aesthetics of impressionist revolution in most striking form. Highly readable classic by great painter. Reproduction of edition designed by Whistler. Introduction by Alfred Werner. xxxvi + 334pp.

21875-9 Paperbound $2.25

"ESSENTIAL GRAMMAR" SERIES

All you really need to know about modern, colloquial grammar.. Many educational shortcuts help you learn faster, understand better. Detailed cognate lists teach you to recognize similarities between English and foreign words and roots—make learning vocabulary easy and interesting. Excellent for independent study or as a supplement to record courses.

ESSENTIAL FRENCH GRAMMAR, Seymour Resnick. 2500-item cognate list. 159pp.
(EBE) 20419-7 Paperbound $1.25

ESSENTIAL GERMAN GRAMMAR, Guy Stern and Everett F. Bleiler. Unusual shortcuts on noun declension, word order, compound verbs. 124pp.
(EBE) 20422-7 Paperbound $1.25

ESSENTIAL ITALIAN GRAMMAR, Olga Ragusa. 111pp.
(EBE) 20779-X Paperbound $1.25

ESSENTIAL JAPANESE GRAMMAR, Everett F. Bleiler. In Romaji transcription; no characters needed. Japanese grammar is regular and simple. 156pp.
21027-8 Paperbound $1.25

ESSENTIAL PORTUGUESE GRAMMAR, Alexander da R. Prista. vi + 114pp.
21650-0 Paperbound $1.25

ESSENTIAL SPANISH GRAMMAR, Seymour Resnick. 2500 word cognate list. 115pp.
(EBE) 20780-3 Paperbound $1.25

ESSENTIAL ENGLISH GRAMMAR, Philip Gucker. Combines best features of modern, functional and traditional approaches. For refresher, class use, home study. x + 177pp.
21649-7 Paperbound $1.25

A PHRASE AND SENTENCE DICTIONARY OF SPOKEN SPANISH. Prepared for U. S. War Department by U. S. linguists. As above, unit is idiom, phrase or sentence rather than word. English-Spanish and Spanish-English sections contain modern equivalents of over 18,000 sentences. Introduction and appendix as above. iv + 513pp.
20495-2 Paperbound $2.00

A PHRASE AND SENTENCE DICTIONARY OF SPOKEN RUSSIAN. Dictionary prepared for U. S. War Department by U. S. linguists. Basic unit is not the word, but the idiom, phrase or sentence. English-Russian and Russian-English sections contain modern equivalents for over 30,000 phrases. Grammatical introduction covers phonetics, writing, syntax. Appendix of word lists for food, numbers, geographical names, etc. vi + 573 pp. 6⅛ x 9¼.
20496-0 Paperbound $3.00

CONVERSATIONAL CHINESE FOR BEGINNERS, Morris Swadesh. Phonetic system, beginner's course in Pai Hua Mandarin Chinese covering most important, most useful speech patterns. Emphasis on modern colloquial usage. Formerly *Chinese in Your Pocket.* xvi + 158pp.
21123-1 Paperbound $1.50

How to Know the Wild Flowers, Mrs. William Starr Dana. This is the classical book of American wildflowers (of the Eastern and Central United States), used by hundreds of thousands. Covers over 500 species, arranged in extremely easy to use color and season groups. Full descriptions, much plant lore. This Dover edition is the fullest ever compiled, with tables of nomenclature changes. 174 full-page plates by M. Satterlee. xii + 418pp. 20332-8 Paperbound $2.50

Our Plant Friends and Foes, William Atherton DuPuy. History, economic importance, essential botanical information and peculiarities of 25 common forms of plant life are provided in this book in an entertaining and charming style. Covers food plants (potatoes, apples, beans, wheat, almonds, bananas, etc.), flowers (lily, tulip, etc.), trees (pine, oak, elm, etc.), weeds, poisonous mushrooms and vines, gourds, citrus fruits, cotton, the cactus family, and much more. 108 illustrations. xiv + 290pp. 22272-1 Paperbound $2.00

How to Know the Ferns, Frances T. Parsons. Classic survey of Eastern and Central ferns, arranged according to clear, simple identification key. Excellent introduction to greatly neglected nature area. 57 illustrations and 42 plates. xvi + 215pp. 20740-4 Paperbound $1.75

Manual of the Trees of North America, Charles S. Sargent. America's foremost dendrologist provides the definitive coverage of North American trees and tree-like shrubs. 717 species fully described and illustrated: exact distribution, down to township; full botanical description; economic importance; description of subspecies and races; habitat, growth data; similar material. Necessary to every serious student of tree-life. Nomenclature revised to present. Over 100 locating keys. 783 illustrations. lii + 934pp. 20277-1, 20278-X Two volumes, Paperbound $6.00

Our Northern Shrubs, Harriet L. Keeler. Fine non-technical reference work identifying more than 225 important shrubs of Eastern and Central United States and Canada. Full text covering botanical description, habitat, plant lore, is paralleled with 205 full-page photographs of flowering or fruiting plants. Nomenclature revised by Edward G. Voss. One of few works concerned with shrubs. 205 plates, 35 drawings. xxviii + 521pp. 21989-5 Paperbound $3.75

The Mushroom Handbook, Louis C. C. Krieger. Still the best popular handbook: full descriptions of 259 species, cross references to another 200. Extremely thorough text enables you to identify, know all about any mushroom you are likely to meet in eastern and central U. S. A.: habitat, luminescence, poisonous qualities, use, folklore, etc. 32 color plates show over 50 mushrooms, also 126 other illustrations. Finding keys. vii + 560pp. 21861-9 Paperbound $3.95

Handbook of Birds of Eastern North America, Frank M. Chapman. Still much the best single-volume guide to the birds of Eastern and Central United States. Very full coverage of 675 species, with descriptions, life habits, distribution, similar data. All descriptions keyed to two-page color chart. With this single volume the average birdwatcher needs no other books. 1931 revised edition. 195 illustrations. xxxvi + 581pp. 21489-3 Paperbound $3.25

AMERICAN FOOD AND GAME FISHES, David S. Jordan and Barton W. Evermann. Definitive source of information, detailed and accurate enough to enable the sportsman and nature lover to identify conclusively some 1,000 species and sub-species of North American fish, sought for food or sport. Coverage of range, physiology, habits, life history, food value. Best methods of capture, interest to the angler, advice on bait, fly-fishing, etc. 338 drawings and photographs. 1 + 574pp. 6⅝ x 9⅜.

22383-1 Paperbound $4.50

THE FROG BOOK, Mary C. Dickerson. Complete with extensive finding keys, over 300 photographs, and an introduction to the general biology of frogs and toads, this is the classic non-technical study of Northeastern and Central species. 58 species; 290 photographs and 16 color plates. xvii + 253pp.

21973-9 Paperbound $4.00

THE MOTH BOOK: A GUIDE TO THE MOTHS OF NORTH AMERICA, William J. Holland. Classical study, eagerly sought after and used for the past 60 years. Clear identification manual to more than 2,000 different moths, largest manual in existence. General information about moths, capturing, mounting, classifying, etc., followed by species by species descriptions. 263 illustrations plus 48 color plates show almost every species, full size. 1968 edition, preface, nomenclature changes by A. E. Brower. xxiv + 479pp. of text. 6½ x 9¼.

21948-8 Paperbound $5.00

THE SEA-BEACH AT EBB-TIDE, Augusta Foote Arnold. Interested amateur can identify hundreds of marine plants and animals on coasts of North America; marine algae; seaweeds; squids; hermit crabs; horse shoe crabs; shrimps; corals; sea anemones; etc. Species descriptions cover: structure; food; reproductive cycle; size; shape; color; habitat; etc. Over 600 drawings. 85 plates. xii + 490pp.

21949-6 Paperbound $3.50

COMMON BIRD SONGS, Donald J. Borror. 33⅓ 12-inch record presents songs of 60 important birds of the eastern United States. A thorough, serious record which provides several examples for each bird, showing different types of song, individual variations, etc. Inestimable identification aid for birdwatcher. 32-page booklet gives text about birds and songs, with illustration for each bird.

21829-5 Record, book, album. Monaural. $2.75

FADS AND FALLACIES IN THE NAME OF SCIENCE, Martin Gardner. Fair, witty appraisal of cranks and quacks of science: Atlantis, Lemuria, hollow earth, flat earth, Velikovsky, orgone energy, Dianetics, flying saucers, Bridey Murphy, food fads, medical fads, perpetual motion, etc. Formerly "In the Name of Science." x + 363pp.

20394-8 Paperbound $2.00

HOAXES, Curtis D. MacDougall. Exhaustive, unbelievably rich account of great hoaxes: Locke's moon hoax, Shakespearean forgeries, sea serpents, Loch Ness monster, Cardiff giant, John Wilkes Booth's mummy, Disumbrationist school of art, dozens more; also journalism, psychology of hoaxing. 54 illustrations. xi + 338pp.

20465-0 Paperbound $2.75

THE PRINCIPLES OF PSYCHOLOGY, William James. The famous long course, complete and unabridged. Stream of thought, time perception, memory, experimental methods—these are only some of the concerns of a work that was years ahead of its time and still valid, interesting, useful. 94 figures. Total of xviii + 1391pp.
20381-6, 20382-4 Two volumes, Paperbound $6.00

THE STRANGE STORY OF THE QUANTUM, Banesh Hoffmann. Non-mathematical but thorough explanation of work of Planck, Einstein, Bohr, Pauli, de Broglie, Schrödinger, Heisenberg, Dirac, Feynman, etc. No technical background needed. "Of books attempting such an account, this is the best," Henry Margenau, Yale. 40-page "Postscript 1959." xii + 285pp.
20518-5 Paperbound $2.00

THE RISE OF THE NEW PHYSICS, A. d'Abro. Most thorough explanation in print of central core of mathematical physics, both classical and modern; from Newton to Dirac and Heisenberg. Both history and exposition; philosophy of science, causality, explanations of higher mathematics, analytical mechanics, electromagnetism, thermodynamics, phase rule, special and general relativity, matrices. No higher mathematics needed to follow exposition, though treatment is elementary to intermediate in level. Recommended to serious student who wishes verbal understanding. 97 illustrations. xvii + 982pp.
20003-5, 20004-3 Two volumes, Paperbound $5.50

GREAT IDEAS OF OPERATIONS RESEARCH, Jagjit Singh. Easily followed non-technical explanation of mathematical tools, aims, results: statistics, linear programming, game theory, queueing theory, Monte Carlo simulation, etc. Uses only elementary mathematics. Many case studies, several analyzed in detail. Clarity, breadth make this excellent for specialist in another field who wishes background. 41 figures. x + 228pp.
21886-4 Paperbound $2.25

GREAT IDEAS OF MODERN MATHEMATICS: THEIR NATURE AND USE, Jagjit Singh. Internationally famous expositor, winner of Unesco's Kalinga Award for science popularization explains verbally such topics as differential equations, matrices, groups, sets, transformations, mathematical logic and other important modern mathematics, as well as use in physics, astrophysics, and similar fields. Superb exposition for layman, scientist in other areas. viii + 312pp.
20587-8 Paperbound $2.25

GREAT IDEAS IN INFORMATION THEORY, LANGUAGE AND CYBERNETICS, Jagjit Singh. The analog and digital computers, how they work, how they are like and unlike the human brain, the men who developed them, their future applications, computer terminology. An essential book for today, even for readers with little math. Some mathematical demonstrations included for more advanced readers. 118 figures. Tables. ix + 338pp.
21694-2 Paperbound $2.25

CHANCE, LUCK AND STATISTICS, Horace C. Levinson. Non-mathematical presentation of fundamentals of probability theory and science of statistics and their applications. Games of chance, betting odds, misuse of statistics, normal and skew distributions, birth rates, stock speculation, insurance. Enlarged edition. Formerly "The Science of Chance." xiii + 357pp.
21007-3 Paperbound $2.00

PLANETS, STARS AND GALAXIES: DESCRIPTIVE ASTRONOMY FOR BEGINNERS, A. E. Fanning. Comprehensive introductory survey of astronomy: the sun, solar system, stars, galaxies, universe, cosmology; up-to-date, including quasars, radio stars, etc. Preface by Prof. Donald Menzel. 24pp. of photographs. 189pp. 5¼ x 8¼.
21680-2 Paperbound $1.50

TEACH YOURSELF CALCULUS, P. Abbott. With a good background in algebra and trig, you can teach yourself calculus with this book. Simple, straightforward introduction to functions of all kinds, integration, differentiation, series, etc. "Students who are beginning to study calculus method will derive great help from this book." Faraday House Journal. 308pp.
20683-1 Clothbound $2.00

TEACH YOURSELF TRIGONOMETRY, P. Abbott. Geometrical foundations, indices and logarithms, ratios, angles, circular measure, etc. are presented in this sound, easy-to-use text. Excellent for the beginner or as a brush up, this text carries the student through the solution of triangles. 204pp.
20682-3 Clothbound $2.00

TEACH YOURSELF ANATOMY, David LeVay. Accurate, inclusive, profusely illustrated account of structure, skeleton, abdomen, muscles, nervous system, glands, brain, reproductive organs, evolution. "Quite the best and most readable account,' Medical Officer. 12 color plates. 164 figures. 311pp. 4¾ x 7.
21651-9 Clothbound $2.50

TEACH YOURSELF PHYSIOLOGY, David LeVay. Anatomical, biochemical bases; digestive, nervous, endocrine systems; metabolism; respiration; muscle; excretion; temperature control; reproduction. "Good elementary exposition," The Lancet. 6 color plates. 44 illustrations. 208pp. 4¼ x 7.
21658-6 Clothbound $2.50

THE FRIENDLY STARS, Martha Evans Martin. Classic has taught naked-eye observation of stars, planets to hundreds of thousands, still not surpassed for charm, lucidity, adequacy. Completely updated by Professor Donald H. Menzel, Harvard Observatory. 25 illustrations. 16 x 30 chart. x + 147pp.
21099-5 Paperbound $1.25

MUSIC OF THE SPHERES: THE MATERIAL UNIVERSE FROM ATOM TO QUASAR, SIMPLY EXPLAINED, Guy Murchie. Extremely broad, brilliantly written popular account begins with the solar system and reaches to dividing line between matter and nonmatter; latest understandings presented with exceptional clarity. Volume One: Planets, stars, galaxies, cosmology, geology, celestial mechanics, latest astronomical discoveries; Volume Two: Matter, atoms, waves, radiation, relativity, chemical action, heat, nuclear energy, quantum theory, music, light, color, probability, antimatter, antigravity, and similar topics. 319 figures. 1967 (second) edition. Total of xx + 644pp.
21809-0, 21810-4 Two volumes, Paperbound $4.00

OLD-TIME SCHOOLS AND SCHOOL BOOKS, Clifton Johnson. Illustrations and rhymes from early primers, abundant quotations from early textbooks, many anecdotes of school life enliven this study of elementary schools from Puritans to middle 19th century. Introduction by Carl Withers. 234 illustrations. xxxiii + 381pp.
21031-6 Paperbound $2.50

THE PHILOSOPHY OF THE UPANISHADS, Paul Deussen. Clear, detailed statement of upanishadic system of thought, generally considered among best available. History of these works, full exposition of system emergent from them, parallel concepts in the West. Translated by A. S. Geden. xiv + 429pp.
21616-0 Paperbound $3.00

LANGUAGE, TRUTH AND LOGIC, Alfred J. Ayer. Famous, remarkably clear introduction to the Vienna and Cambridge schools of Logical Positivism; function of philosophy, elimination of metaphysical thought, nature of analysis, similar topics. "Wish I had written it myself," Bertrand Russell. 2nd, 1946 edition. 160pp.
20010-8 Paperbound $1.35

THE GUIDE FOR THE PERPLEXED, Moses Maimonides. Great classic of medieval Judaism, major attempt to reconcile revealed religion (Pentateuch, commentaries) and Aristotelian philosophy. Enormously important in all Western thought. Unabridged Friedländer translation. 50-page introduction. lix + 414pp.
(USO) 20351-4 Paperbound $2.50

OCCULT AND SUPERNATURAL PHENOMENA, D. H. Rawcliffe. Full, serious study of the most persistent delusions of mankind: crystal gazing, mediumistic trance, stigmata, lycanthropy, fire walking, dowsing, telepathy, ghosts, ESP, etc., and their relation to common forms of abnormal psychology. Formerly *Illusions and Delusions of the Supernatural and the Occult.* iii + 551pp.   20503-7 Paperbound $3.50

THE EGYPTIAN BOOK OF THE DEAD: THE PAPYRUS OF ANI, E. A. Wallis Budge. Full hieroglyphic text, interlinear transliteration of sounds, word for word translation, then smooth, connected translation; Theban recension. Basic work in Ancient Egyptian civilization; now even more significant than ever for historical importance, dilation of consciousness, etc. clvi + 377pp. 6½ x 9¼.
21866-X Paperbound $3.75

PSYCHOLOGY OF MUSIC, Carl E. Seashore. Basic, thorough survey of everything known about psychology of music up to 1940's; essential reading for psychologists, musicologists. Physical acoustics; auditory apparatus; relationship of physical sound to perceived sound; role of the mind in sorting, altering, suppressing, creating sound sensations; musical learning, testing for ability, absolute pitch, other topics. Records of Caruso, Menuhin analyzed. 88 figures. xix + 408pp.
21851-1 Paperbound $2.75

THE I CHING (THE BOOK OF CHANGES), translated by James Legge. Complete translated text plus appendices by Confucius, of perhaps the most penetrating divination book ever compiled. Indispensable to all study of early Oriental civilizations. 3 plates. xxiii + 448pp.   21062-6 Paperbound $2.75

THE UPANISHADS, translated by Max Müller. Twelve classical upanishads: Chandogya, Kena, Aitareya, Kaushitaki, Isa, Katha, Mundaka, Taittiriyaka, Brhadaranyaka, Svetasvatara, Prasna, Maitriyana. 160-page introduction, analysis by Prof. Müller. Total of 826pp.   20398-0, 20399-9 Two volumes, Paperbound $5.00

CATALOGUE OF DOVER BOOKS

MATHEMATICAL PUZZLES FOR BEGINNERS AND ENTHUSIASTS, Geoffrey Mott-Smith. 189 puzzles from easy to difficult—involving arithmetic, logic, algebra, properties of digits, probability, etc.—for enjoyment and mental stimulus. Explanation of mathematical principles behind the puzzles. 135 illustrations. viii + 248pp.
20198-8 Paperbound $1.25

PAPER FOLDING FOR BEGINNERS, William D. Murray and Francis J. Rigney. Easiest book on the market, clearest instructions on making interesting, beautiful origami. Sail boats, cups, roosters, frogs that move legs, bonbon boxes, standing birds, etc. 40 projects; more than 275 diagrams and photographs. 94pp.
20713-7 Paperbound $1.00

TRICKS AND GAMES ON THE POOL TABLE, Fred Herrmann. 79 tricks and games—some solitaires, some for two or more players, some competitive games—to entertain you between formal games. Mystifying shots and throws, unusual caroms, tricks involving such props as cork, coins, a hat, etc. Formerly *Fun on the Pool Table.* 77 figures. 95pp.
21814-7 Paperbound $1.00

HAND SHADOWS TO BE THROWN UPON THE WALL: A SERIES OF NOVEL AND AMUSING FIGURES FORMED BY THE HAND, Henry Bursill. Delightful picturebook from great-grandfather's day shows how to make 18 different hand shadows: a bird that flies, duck that quacks, dog that wags his tail, camel, goose, deer, boy, turtle, etc. Only book of its sort. vi + 33pp. 6½ x 9¼. 21779-5 Paperbound $1.00

WHITTLING AND WOODCARVING, E. J. Tangerman. 18th printing of best book on market. "If you can cut a potato you can carve" toys and puzzles, chains, chessmen, caricatures, masks, frames, woodcut blocks, surface patterns, much more. Information on tools, woods, techniques. Also goes into serious wood sculpture from Middle Ages to present, East and West. 464 photos, figures. x + 293pp.
20965-2 Paperbound $2.00

HISTORY OF PHILOSOPHY, Julián Marías. Possibly the clearest, most easily followed, best planned, most useful one-volume history of philosophy on the market; neither skimpy nor overfull. Full details on system of every major philosopher and dozens of less important thinkers from pre-Socratics up to Existentialism and later. Strong on many European figures usually omitted. Has gone through dozens of editions in Europe. 1966 edition, translated by Stanley Appelbaum and Clarence Strowbridge. xviii + 505pp.
21739-6 Paperbound $2.75

YOGA: A SCIENTIFIC EVALUATION, Kovoor T. Behanan. Scientific but non-technical study of physiological results of yoga exercises; done under auspices of Yale U. Relations to Indian thought, to psychoanalysis, etc. 16 photos. xxiii + 270pp.
20505-3 Paperbound $2.50

Prices subject to change without notice.
Available at your book dealer or write for free catalogue to Dept. GI, Dover Publications, Inc., 180 Varick St., N. Y., N. Y. 10014. Dover publishes more than 150 books each year on science, elementary and advanced mathematics, biology, music, art, literary history, social sciences and other areas.